SHIRLEY DISSELER'S STRATEGIES AND ACTIVITIES FOR COMMON CORE MATH

GRADES 3–5

PART 1

SHIRLEY DISSELER'S STRATEGIES AND ACTIVITIES FOR COMMON CORE MATH

GRADES 3–5

PART 1

✓ Operations & Algebraic Thinking

✓ Numbers & Operations in Base Ten

✓ Numbers & Operations—Fractions

COMPASS

Shirley Disseler's Strategies & Activities for Common Core Math: Grades 3–5, Part 1

Copyright © 2014 by Shirley Disseler
Published by Brigantine Media/Compass Publishing
211 North Avenue, St. Johnsbury, Vermont 05819

Cover and book design by Jacob L. Grant

All rights reserved.
Your individual purchase of this book entitles you to reproduce these pages as needed for your own classroom use only. Otherwise, no part of this book may be reproduced or utilized in any way or by any means, electronic or mechanical, including photocopying, recording, or information storage or retrieval system, without prior written permission from the publisher. Individual copies may not be distributed in any other form.

Brigantine Media/Compass Publishing
211 North Avenue
St. Johnsbury, Vermont 05819
Phone: 802-751-8802
Fax: 802-751-8804
E-mail: neil@brigantinemedia.com
Website: www.compasspublishing.org

ORDERING INFORMATION

Quantity sales
Special discounts for schools are available for quantity purchases of physical books and digital downloads. For information, contact Brigantine Media at the address shown above or visit www.compasspublishing.org.

Individual sales
Brigantine Media/Compass Publishing publications are available through most bookstores. They can also be ordered directly from Brigantine Media.
Phone: 802-751-8802 | Fax: 802-751-8804
www.compasspublishing.org

ISBN 978-1-9384063-6-2

DEDICATION

This book is dedicated to my husband Tom, and my two sons, Ryan and Steven. They have provided inspiration to me throughout the years. This inspiration has created within me a need to share my ideas with others in the field of education.

In honor and remembrance of:
Miss Christine Dunn: The reason I am a teacher today.
Mrs. Betty Dunn: My mom, whom I miss every day.

ACKNOWLEDGMENTS

I would like to acknowledge the following persons for their support in this project:

This book would not be possible without the encouragement of my husband, Tom Disseler.

My son Ryan Disseler, also a teacher, for his constant listening and critiquing of my ideas.

My dad, who has always taught me to go the extra mile and reach for the stars.

My former teammate Angela London, with whom I have shared all of my math ideas throughout teaching middle school.

My students at High Point University, who have used my activities and provided feedback, participated in the activities in class, and listened to my ideas relentlessly.

Mariann Tillery, the dean of the School of Education at High Point University, for her endless support in this endeavor.

My publisher and editor, Neil Raphel and Janis Raye, for their continuous e-mails and hard work in getting this to press.

Laura Candler, for helping me get started with publication efforts.

My colleagues at High Point University, who have always given me support in times of stress.

To teachers everywhere! This is for you.

CONTENTS

Introduction..1
Common Core Standards Chart..................................4
Standards for Mathematical Practice Chart..............10

OPERATIONS & ALGEBRAIC THINKING..............11

    Equal Shares: Multiplication....................................13
    Equal Shares: Division..30
    Resolve to Solve...32
    Mystery Numbers...36
    Distributive Property Match-Up..............................41
    Speed Sets...43
    Modeling Properties..51
    Order Up...55
    Word Problem Frenzy..57
    Venn Number Theory..65
    Visual Patterning...69
    Comparative Equations..72
    Prime Circles..83
    Prime or Composite?...86
    Hit the Target...90
    The Number Is in the Word....................................93

NUMBERS & OPERATIONS IN BASE TEN..............97

    Who's in Whose Place?..98
    Card Sharks..103
    Race to the Place..106
    Area Model Match-Up..109
    Number Logic in a Box...115
    Place the Number..120

CONTENTS

NUMBERS & OPERATIONS—FRACTIONS 127

    Play Your Card 128

    Equivalent Fraction Frenzy 133

    Fractions Rule! 138

    Meet Me in the Middle 141

    Where Do I Belong? 145

    Make a Buck 151

    Rolling into Fractions 157

    How Does Your Garden Grow? 159

    The Great Fraction Hunt 165

    Decimals and Fractions in Action 168

    Name That Decimal 176

    Featuring Fractions 178

Bibliography 183

INTRODUCTION

Ask your students: "When do you use math outside the classroom?" Some will reply, "Never." These students are not able to link what they learn in math class to their own lives or to the world around them. Often these are the students with the greatest math deficits. Other students will answer that math is used when you buy something or measure something. That's a start. These students recognize some degree of association of math to their day-to-day activities.

The focus of the Common Core Standards for Mathematics is for students to learn to understand math on a deep level. Students can't simply memorize algorithms to solve math problems; they need to understand why those computational methods work. Math teachers need to make sure students can use numbers in real-world applications.

The Common Core Standards for Mathematics require a creative approach to teaching. Teachers are expected to teach students how to think critically about mathematics. But to do this, teachers need a new set of strategies and activities to get and keep kids engaged in mathematics.

Shirley Disseler's *Strategies and Activities for Common Core Math: Grades 3 – 5* provides dozens of activities that align with the Common Core Standards for grades 3 to 5 and the Standards for Mathematical Practice, enhance the curriculum, and create a sense of "fun" in the classroom. Each activity has been tested by children in grades 3 to 5, and they work.

HOW TO USE THIS PROGRAM

This program is divided into two books for ease of use, with six chapters in all. Five chapters align with the five domains in the Common Core Math Standards for grades 3 through 5. This book, Part 1, covers the domains of Operations & Algebraic Thinking, Number & Operations in Base Ten, and Number & Operations—Fractions. The last chapter in Part 2 offers enrichment activities that promote even more critical thinking using mathematics. These kinds of activities can help to get the mathematical conversation started in the classroom, and often serve to engage the most reluctant learners.

PART 1
Operations & Algebraic Thinking
Number & Operations in Base Ten
Number & Operations—Fractions

PART 2
Measurement & Data
Geometry
Mathematical Practices Across the Common Core

The activities in Parts 1 and 2 help to teach every Common Core Standard for Mathematics in grades 3 through 5. A chart that shows each standard and the activity or activities that align to that standard starts on page 4. When teachers plan lessons to work on specific Common Core Standards, they can check the chart, find the appropriate activities, and incorporate those activities into their instruction and assessment. For each Common Core domain, there are a number of activities that a teacher can use to teach one or more standards. All the materials needed for an activity are listed with it, and all the game boards, cards, recording sheets, and assessments for the activities are included in the book.

Every activity also aligns with one or more of the Common Core Standards for Mathematical Practice. These are as integral to mathematics learning as the content standards. The Standards for Mathematical Practice describe the kind of process expertise that is developed with the study of math, including: making sense of problems and persevering in their solution; being able to reason abstractly and quantitatively; learning to model with mathematics and use tools strategically, and more. A chart that explains the importance of the Standards for Mathematical Practices for students in grades 3 through 5 is on page 10.

CHARTS

Common Core Standards for Mathematics, Grades 3–5 / Teaching Activities
- Operations & Algebraic Thinking
- Number & Operations in Base Ten
- Number & Operations—Fractions

Standards for Mathematical Practice / Implications in Grades 3–5

© Copyright 2010 National Governors Association Center for Best Practices and Council of Chief State School Officers. All rights reserved.

Operations & Algebraic Thinking

COMMON CORE STANDARDS	ACTIVITIES
Grade 3	
3.OA.A.1 Interpret products of whole numbers, e.g., interpret 5 × 7 as the total number of objects in 5 groups of 7 objects each. For example, describe a context in which a total number of objects can be expressed as 5 × 7.	Equal Shares—Multiplication (p. 13)
3.OA.A.2 Interpret whole-number quotients of whole numbers, e.g., interpret 56 ÷ 8 as the number of objects in each share when 56 objects are partitioned equally into 8 shares, or as a number of shares when 56 objects are partitioned into equal shares of 8 objects each.	Equal Shares—Division (p. 30)
3.OA.A.3 Use multiplication and division within 100 to solve word problems in situations involving equal groups, arrays, and measurement quantities, e.g., by using drawings and equations with a symbol for the unknown number to represent the problem.	Resolve to Solve (p. 32)
3.OA.A.4 Determine the unknown whole number in a multiplication or division equation relating three whole numbers.	Mystery Numbers (p. 36)
3.OA.B.5 Apply properties of operations as strategies to multiply and divide.	Distributive Property Match-Up (p. 41) Speed Sets (p. 43)
3.OA.B.6 Understand division as an unknown-factor problem.	Mystery Numbers (p. 36) Modeling Properties (p. 51)
3.OA.C.7 Fluently multiply and divide within 100, using strategies such as the relationship between multiplication and division (e.g., knowing that 8 × 5 = 40, one knows 40 ÷ 5 = 8) or properties of operations. By the end of Grade 3, know from memory all products of two one-digit numbers.	Order Up (p. 55)
3.OA.D.8 Solve two-step word problems using the four operations. Represent these problems using equations with a letter standing for the unknown quantity. Assess the reasonableness of answers using mental computation and estimation strategies including rounding.	Word Problem Frenzy (p. 57)
3.OA.D.9 Identify arithmetic patterns (including patterns in the addition table or multiplication table), and explain them using properties of operations.	Venn Number Theory (p. 65) Visual Patterning (p. 69)
Grade 4	
4.OA.A.1 Interpret a multiplication equation as a comparison, e.g., interpret 35 = 5 × 7 as a statement that 35 is 5 times as many as 7 and 7 times as many as 5. Represent verbal statements of multiplicative comparisons as multiplication equations.	Comparative Equations (p. 72)
4.OA.A.2 Multiply or divide to solve word problems involving multiplicative comparison, e.g., by using drawings and equations with a symbol for the unknown number to represent the problem, distinguishing multiplicative comparison from additive comparison.	Comparative Equations (p. 72)
4.OA.A.3 Solve multistep word problems posed with whole numbers and having whole-number answers using the four operations, including problems in which remainders must be interpreted. Represent these problems using equations with a letter standing for the unknown quantity. Assess the reasonableness of answers using mental computation and estimation strategies including rounding.	Word Problem Frenzy (p. 57)

COMMON CORE STANDARDS	ACTIVITIES
4.OA.B.4 Find all factor pairs for a whole number in the range 1-100. Recognize that a whole number is a multiple of each of its factors. Determine whether a given whole number in the range 1-100 is a multiple of a given one-digit number. Determine whether a given whole number in the range 1-100 is prime or composite.	Venn Number Theory (p. 65) Prime Circles (p. 83) Prime or Composite? (p. 86) Hit the Target (p. 90)
4.OA.C.5 Generate a number or shape pattern that follows a given rule. Identify apparent features of the pattern that were not explicit in the rule itself.	Number Logic in a Box (p. 115)

Grade 5

COMMON CORE STANDARDS	ACTIVITIES
5.OA.A.1 Use parentheses, brackets, or braces in numerical expressions, and evaluate expressions with these symbols.	Distributive Property Match-Up (p. 41) Order Up (p. 55)
5.OA.A.2 Write simple expressions that record calculations with numbers, and interpret numerical expressions without evaluating them.	The Number Is in the Word (p. 93)
5.OA.B.3 Generate two numerical patterns using two given rules. Identify apparent relationships between corresponding terms. Form ordered pairs consisting of corresponding terms from the two patterns, and graph the ordered pairs on a coordinate plane.	Venn Number Theory (p. 65)

Numbers & Operations In Base Ten

COMMON CORE STANDARDS	ACTIVITIES
Grade 3	
3.NBT.A.1 Use place value understanding to round whole numbers to the nearest 10 or 100.	Who's in Whose Place? (p. 98)
3.NBT.A.2 Fluently add and subtract within 1000 using strategies and algorithms based on place value, properties of operations, and/or the relationship between addition and subtraction.	Order Up (p. 55)
3.NBT.A.3 Multiply one-digit whole numbers by multiples of 10 in the range 10-90 (e.g., 9×80, 5×60) using strategies based on place value and properties of operations.	Card Sharks (p. 103)
Grade 4	
4.NBT.A.1 Recognize that in a multi-digit whole number, a digit in one place represents ten times what it represents in the place to its right.	Who's in Whose Place? (p. 98)
4.NBT.A.2 Read and write multi-digit whole numbers using base-ten numerals, number names, and expanded form. Compare two multi-digit numbers based on meanings of the digits in each place, using >, =, and < symbols to record the results of comparisons.	Who's in Whose Place? (p. 98) Race to the Place (p. 106) Area Model Match-Up (p. 109)
4.NBT.A.3 Use place value understanding to round multi-digit whole numbers to any place.	Race to the Place (p. 106)
4.NBT.B.4 Fluently add and subtract multi-digit whole numbers using the standard algorithm.	Number Logic in a Box (p. 115)
4.NBT.B.5 Multiply a whole number of up to four digits by a one-digit whole number, and multiply two two-digit numbers, using strategies based on place value and the properties of operations. Illustrate and explain the calculation by using equations, rectangular arrays, and/or area models.	Modeling Properties (p. 51) Area Model Match-Up (p. 109)
4.NBT.B.6 Find whole-number quotients and remainders with up to four-digit dividends and one-digit divisors, using strategies based on place value, the properties of operations, and/or the relationship between multiplication and division. Illustrate and explain the calculation by using equations, rectangular arrays, and/or area models.	Area Model Match-Up (p. 109)
Grade 5	
5.NBT.A.1 Recognize that in a multi-digit number, a digit in one place represents 10 times as much as it represents in the place to its right and 1/10 of what it represents in the place to its left.	How Does Your Garden Grow? (p. 159) Place the Number (p. 120)
5.NBT.A.2 Explain patterns in the number of zeros of the product when multiplying a number by powers of 10, and explain patterns in the placement of the decimal point when a decimal is multiplied or divided by a power of 10. Use whole-number exponents to denote powers of 10.	Who's in Whose Place? (p. 98) Place the Number (p. 120)
5.NBT.A.3 Read, write, and compare decimals to thousandths.	Who's in Whose Place? (p. 98) Place the Number (p. 120)
5.NBT.A.4 Use place value understanding to round decimals to any place.	Who's in Whose Place? (p. 98)

COMMON CORE STANDARDS	ACTIVITIES
5.NBT.B.5 Fluently multiply multi-digit whole numbers using the standard algorithm.	Number Logic in a Box (p. 115)
5.NBT.B.6 Find whole-number quotients of whole numbers with up to four-digit dividends and two-digit divisors, using strategies based on place value, the properties of operations, and/or the relationship between multiplication and division. Illustrate and explain the calculation by using equations, rectangular arrays, and/or area models.	Number Logic in a Box (p. 115)
5.NBT.B.7 Add, subtract, multiply, and divide decimals to hundredths, using concrete models or drawings and strategies based on place value, properties of operations, and/or the relationship between addition and subtraction; relate the strategy to a written method and explain the reasoning used.	Place the Number (p. 120)

Number & Operations—Fractions

COMMON CORE STANDARDS	ACTIVITIES
Grade 3	
3.NF.A.1 Understand a fraction 1/b as the quantity formed by 1 part when a whole is partitioned into b equal parts; understand a fraction a/b as the quantity formed by a parts of size 1/b.	Play Your Card (p. 128) Equivalent Fraction Frenzy (p. 133) Fractions Rule! (p. 138)
3.NF.A.2 Understand a fraction as a number on the number line; represent fractions on a number line diagram.	Play Your Card (p. 128) Equivalent Fraction Frenzy (p. 133)
3.NF.A.3 Explain equivalence of fractions in special cases, and compare fractions by reasoning about their size.	Fractions Rule! (p. 138) Meet Me in the Middle (p. 141)
Grade 4	
4.NF.A.1 Explain why a fraction a/b is equivalent to a fraction (n × a)/(n × b) by using visual fraction models, with attention to how the number and size of the parts differ even though the two fractions themselves are the same size. Use this principle to recognize and generate equivalent fractions.	Play Your Card (p. 128) Equivalent Fraction Frenzy (p. 133) Fractions Rule! (p. 138) Meet Me in the Middle (p. 141) Where Do I Belong? (p. 145)
4.NF.A.2 Compare two fractions with different numerators and different denominators, e.g., by creating common denominators or numerators, or by comparing to a benchmark fraction such as 1/2. Recognize that comparisons are valid only when the two fractions refer to the same whole. Record the results of comparisons with symbols >, =, or <, and justify the conclusions, e.g., by using a visual fraction model.	Play Your Card (p. 128) Equivalent Fraction Frenzy (p. 133) Fractions Rule! (p. 138) Meet Me in the Middle (p. 141) Where Do I Belong? (p. 145)
4.NF.B.3 Understand a fraction a/b with a > 1 as a sum of fractions 1/b.	Make a Buck (p. 151) Rolling into Fractions (p. 157) How Does Your Garden Grow? (p. 159) The Great Fraction Hunt (p. 165)
4.NF.B.4 Apply and extend previous understandings of multiplication to multiply a fraction by a whole number.	The Great Fraction Hunt (p. 165)
4.NF.C.5 Express a fraction with denominator 10 as an equivalent fraction with denominator 100, and use this technique to add two fractions with respective denominators 10 and 100.2	How Does Your Garden Grow? (p. 159)
4.NF.C.6 Use decimal notation for fractions with denominators 10 or 100. For example, rewrite 0.62 as 62/100; describe a length as 0.62 meters; locate 0.62 on a number line diagram.	How Does Your Garden Grow? (p. 159) Decimals and Fractions in Action (p. 168) Name That Decimal (p. 176)
4.NF.C.7 Compare two decimals to hundredths by reasoning about their size. Recognize that comparisons are valid only when the two decimals refer to the same whole. Record the results of comparisons with the symbols >, =, or <, and justify the conclusions, e.g., by using a visual model.	How Does Your Garden Grow? (p. 159) Name That Decimal (p. 176)
Grade 5	
5.NF.A.1 Add and subtract fractions with unlike denominators (including mixed numbers) by replacing given fractions with equivalent fractions in such a way as to produce an equivalent sum or difference of fractions with like denominators.	Fractions Rule! (p. 138) Meet Me in the Middle (p. 141) Make a Buck (p. 151) Rolling into Fractions (p. 157) How Does Your Garden Grow? (p. 159) The Great Fraction Hunt (p. 165)

Number & Operations—Fractions

COMMON CORE STANDARDS	ACTIVITIES
5.NF.A.2 Solve word problems involving addition and subtraction of fractions referring to the same whole, including cases of unlike denominators, e.g., by using visual fraction models or equations to represent the problem. Use benchmark fractions and number sense of fractions to estimate mentally and assess the reasonableness of answers.	Fractions Rule! (p. 138) How Does Your Garden Grow? (p. 159)
5.NF.B.3 Interpret a fraction as division of the numerator by the denominator (a/b = a ÷ b). Solve word problems involving division of whole numbers leading to answers in the form of fractions or mixed numbers, e.g., by using visual fraction models or equations to represent the problem.	Rolling into Fractions (p. 157) The Great Fraction Hunt (p. 165)
5.NF.B.4 Apply and extend previous understandings of multiplication to multiply a fraction or whole number by a fraction.	Rolling into Fractions (p. 157) How Does Your Garden Grow? (p. 159)
5.NF.B.5 Interpret multiplication as scaling (resizing), by:	Featuring Fractions (p. 178)
5.NF.B.6 Solve real world problems involving multiplication of fractions and mixed numbers, e.g., by using visual fraction models or equations to represent the problem.	Featuring Fractions (p. 178)
5.NF.B.7 Apply and extend previous understandings of division to divide unit fractions by whole numbers and whole numbers by unit fractions.	The Great Fraction Hunt (p. 165)

STANDARDS FOR MATHEMATICAL PRACTICE	IMPLICATIONS IN GRADES 3 – 5
MP1 Make sense of problems and persevere in solving them.	Students should talk themselves through problems and try various strategies as well as listen to the strategies of others.
MP2 Reason abstractly and quantitatively.	Students should be able to recognize the numerical value of digits and know that they represent a specific amount. Students should be able to connect mathematical symbols to the meaning of the numbers in a problem. By grade three, students should begin to take the concepts of numeration and extend them onward toward fractions and decimals. Place value understanding is key.
MP3 Construct viable arguments and critique the reasoning of others.	Students should be able to use objects and pictures to prove their answers and argue the validity of their answers. Students should also be able to critique the work of others using objects and pictures, explaining corrections needed.
MP4 Model with mathematics.	Students should experiment with math problems in various ways using objects, pictures, charts, lists, graphs, and equations to model the problems. Students should be able to create and understand different representations of the same problem.
MP5 Use appropriate tools strategically.	Students should be able to choose the most appropriate tool and use those tools to solve problems. Tools at this level include (but are not limited to) estimation, graph paper, protractors, and number lines.
MP6 Attend to precision.	Student should use clear and exact math vocabulary when discussing problems with others. Students should be able to label graphs appropriately, use units carefully and precisely, and state the meaning of the math they use to solve a problem.
MP7 Look for and make use of structure.	Students should be able to look for and find patterns in math problems in order to solve them. Example: looking for place value in partial products.
MP8 Look for and express regularity in repeated reasoning.	If there is a repeating action occurring in the solving of a problem, students should be able to notice this and relate it to the appropriate skill. Example: noticing that repeatedly adding the same number indicates the same outcome as multiplying. Using models to generate patterns is important in this practice.

OPERATIONS & ALGEBRAIC THINKING

The Common Core domain of Operations & Algebraic Thinking for grades 3 through 5 emphasizes the conceptual understanding of operations. It includes properties, operational problem-solving, and builds a foundation for algebra through patterns, expressions, and equations.

According to Marilyn Burns (2007), "arithmetic instruction has traditionally centered on developing students' proficiency with paper and pencil computation." Burns notes some common errors in computational mathematics students make. She contends that these are not random, but rather, are consistent and rule-bound. They show that students' procedural mathematics understanding is weak. The following chart shows some of these errors, along with possible reasons for the errors.

COMMON ERRORS IN COMPUTATIONAL MATHEMATICS

Problem Type	Possible Reason for Error
$3 + \underline{\quad} = 12$ (Many students put 15 in the blank)	Students know the sign means to add, so they add the numbers they see.
45 +27 612	Students add the numbers in each column and write the answer below the line.
32 - 25 13	Students take the smaller number from the larger number.
9 9 5̶0̶0 -156 343	Students change the zeros to nines.
1 1 5̶00 -156 354	Students borrow from the 5.

34 $12\overline{)4080}$	Students drop the zero at the end of problems.
½ + ⅔ = ⅗	Students add across the fractions.
$3.0\ 8$ $2.\ 3$ $\underline{+3\ 6}$ $3.6\ 7$	Students line up the numbers and add.
$\$6.20$ $\underline{\times .15}$ 93.00	Students multiply and then bring down the decimal point.

These errors in computation make it clear that students are relying on the process, not the reasoning, when performing operational skills. They show that too often the focus in elementary school has been on procedural knowledge rather than critical thinking.

Teachers need different strategies to teach students algebraic thinking. They must teach students how to create viable arguments that defend their thought processes to help them become critical thinkers in mathematics.

This chapter includes a variety of activity-based ideas to help students master the Common Core domain of Operations & Algebraic Thinking. The activities are designed to promote algebraic thinking about operations in mathematics. Many are games and use simple materials. The activities offer a "hands-on, minds-on" approach to learning and can be used for pre-assessment, enrichment, assessment, or remediation.

EQUAL SHARES: MULTIPLICATION

GRADE 3

COMMON CORE STANDARD
3.OA.A.1

MATHEMATICAL PRACTICES
MP1
MP2

MATERIALS
- Equal Shares playing cards (remove division sentence cards for this game)
- Equal Shares: Multiplication recording sheet (one per student)

OVERVIEW

This game involves identifying groups of objects in a multiplication problem. Common Core requires that students be able to **interpret groups or sets of objects as they refer to a multiplication sentence.**

PROCEDURE

Game is played in pairs

1. Players shuffle the Equal Shares playing cards and deal six cards to each player. The remaining cards are placed face down as a draw pile.
2. Players review the cards they have been dealt to see if they can make a set of three, consisting of one number sentence, one picture card, and one solution card.
3. Players take turns, beginning with the youngest player.
4. If player 1 can make a set, he/she places the set down on the table and records the number sentence, picture, and solution on the recording sheet.
5. If no set can be made, the player discards one card to start a discard pile and draws a card from the draw pile.
6. Play continues until all cards are used.
7. Players count their sets. Winner has the greatest number of sets.

Equal Shares
PLAYING CARDS

18 ÷ 3	(three circles each containing 6 dots)
6 x 3	18
6	5

14   Strategies and Activities for Common Core Math: Grades 3–5   Part 1   Shirley Disseler

Equal Shares
PLAYING CARDS

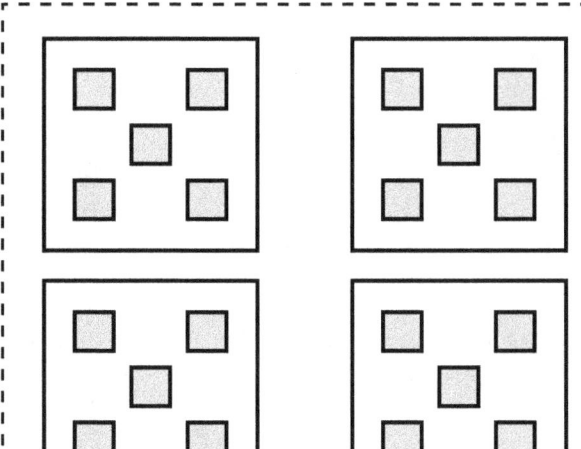	$20 \div 4$
20	4×5
3	$21 \div 7$

Shirley Disseler   Strategies and Activities for Common Core Math: Grades 3–5   Part 1   **15**

Equal Shares
PLAYING CARDS

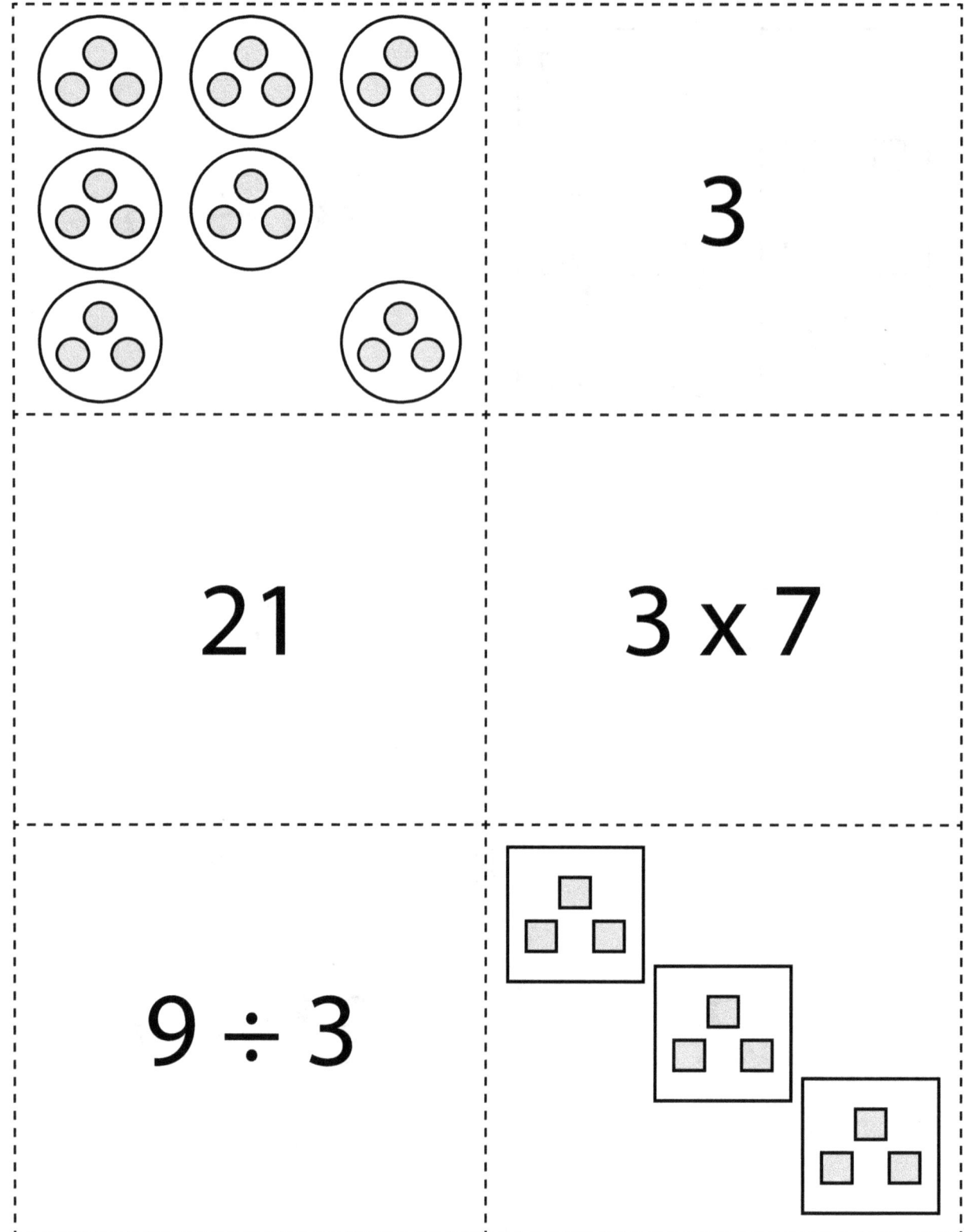

Equal Shares
PLAYING CARDS

3 x 3	9
12 ÷ 4	4 x 3
3	12

Equal Shares
PLAYING CARDS

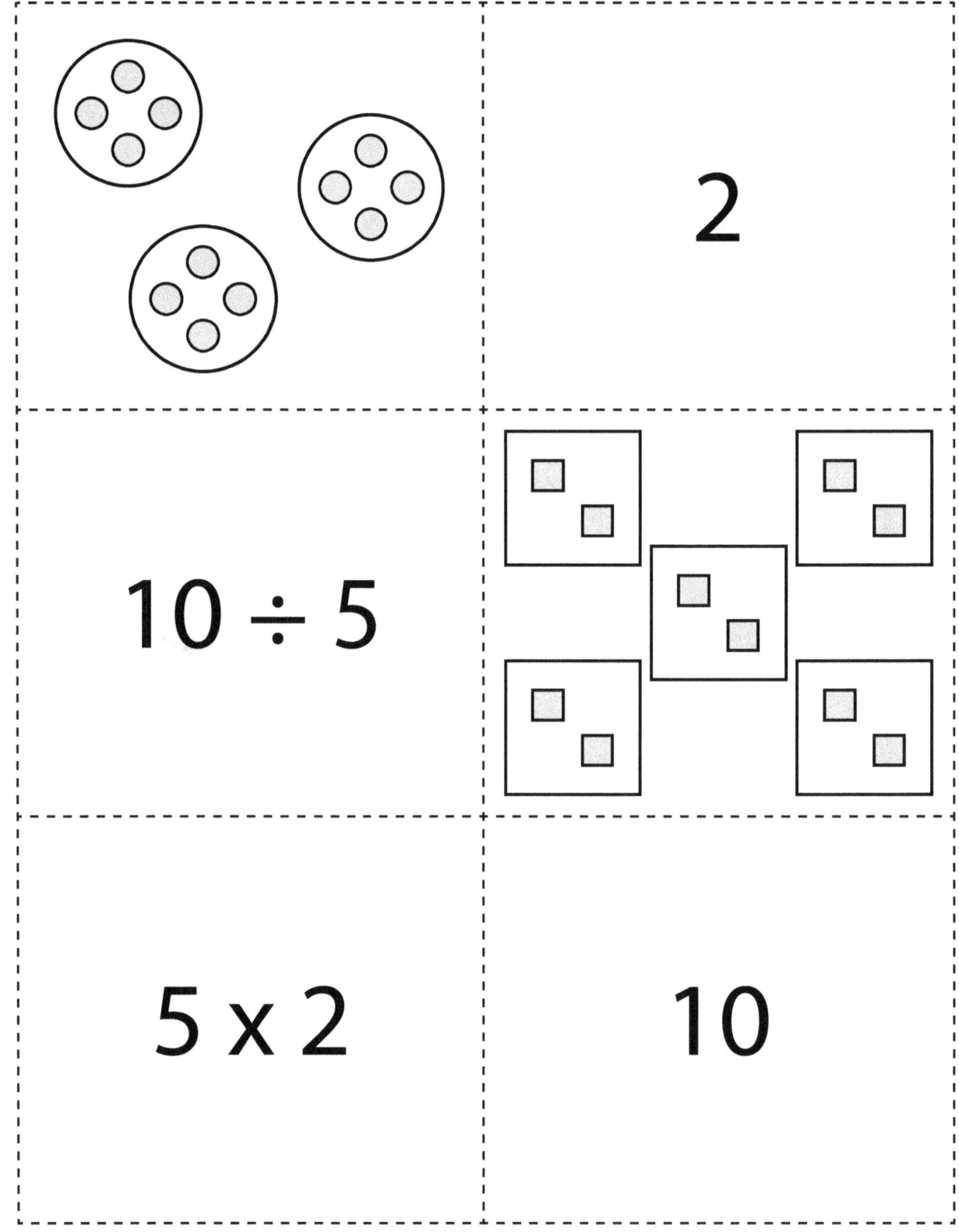

18 Strategies and Activities for Common Core Math: Grades 3–5   Part 1   Shirley Disseler

Equal Shares
PLAYING CARDS

7	
14 ÷ 2	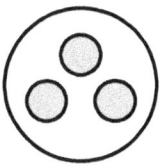 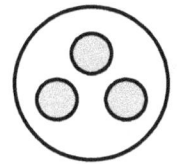 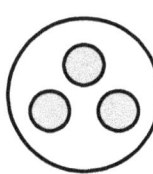 
2 x 7	14

Shirley Disseler   Strategies and Activities for Common Core Math: Grades 3–5   Part 1   **19**

Equal Shares
PLAYING CARDS

3	15
15 ÷ 3	8
3 x 5	24 ÷ 3

Equal Shares
PLAYING CARDS

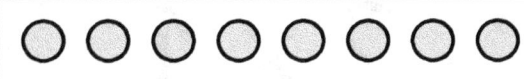	4
3 x 8	24
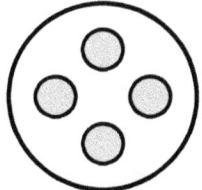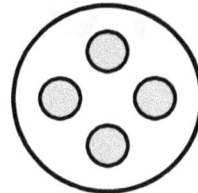	8 ÷ 2

Shirley Disseler — Strategies and Activities for Common Core Math: Grades 3–5 — Part 1

Equal Shares
PLAYING CARDS

8	2 x 4
16 ÷ 2	○○○○○○○○ ○○○○○○○○
2 x 8	16

22   Strategies and Activities for Common Core Math: Grades 3–5   Part 1   Shirley Disseler

Equal Shares
PLAYING CARDS

8	9
$18 \div 2$	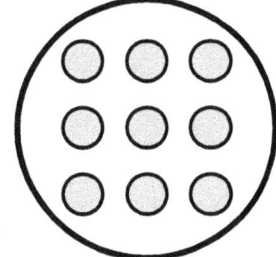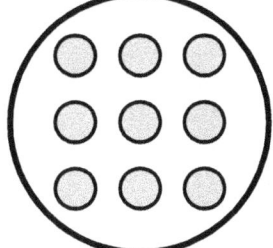
2 x 9	18

Shirley Disseler — Strategies and Activities for Common Core Math: Grades 3–5 — Part 1 — 23

Equal Shares
PLAYING CARDS

6	12 ÷ 2
2	12 ÷ 6
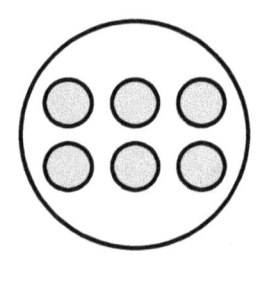	2 x 6

24 Strategies and Activities for Common Core Math: Grades 3–5   Part 1   Shirley Disseler

Equal Shares
PLAYING CARDS

12	▫▫ ▫▫ ▫▫ ▫▫ ▫▫ ▫▫
3 × 9	3
27 ÷ 9	9

Shirley Disseler — Strategies and Activities for Common Core Math: Grades 3–5

Equal Shares
PLAYING CARDS

$27 \div 3$	27
3×3	  
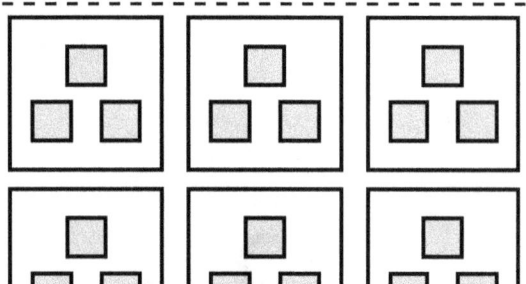 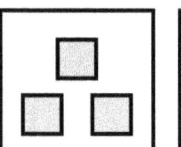 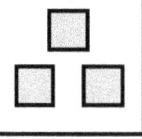	5

Equal Shares
PLAYING CARDS

$10 \div 2$	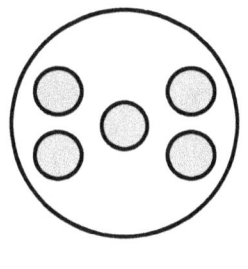 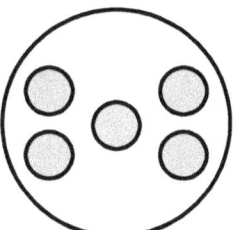
4	2×5
10	$20 \div 5$

Shirley Disseler   Strategies and Activities for Common Core Math: Grades 3–5   Part 1   **27**

Equal Shares
PLAYING CARDS

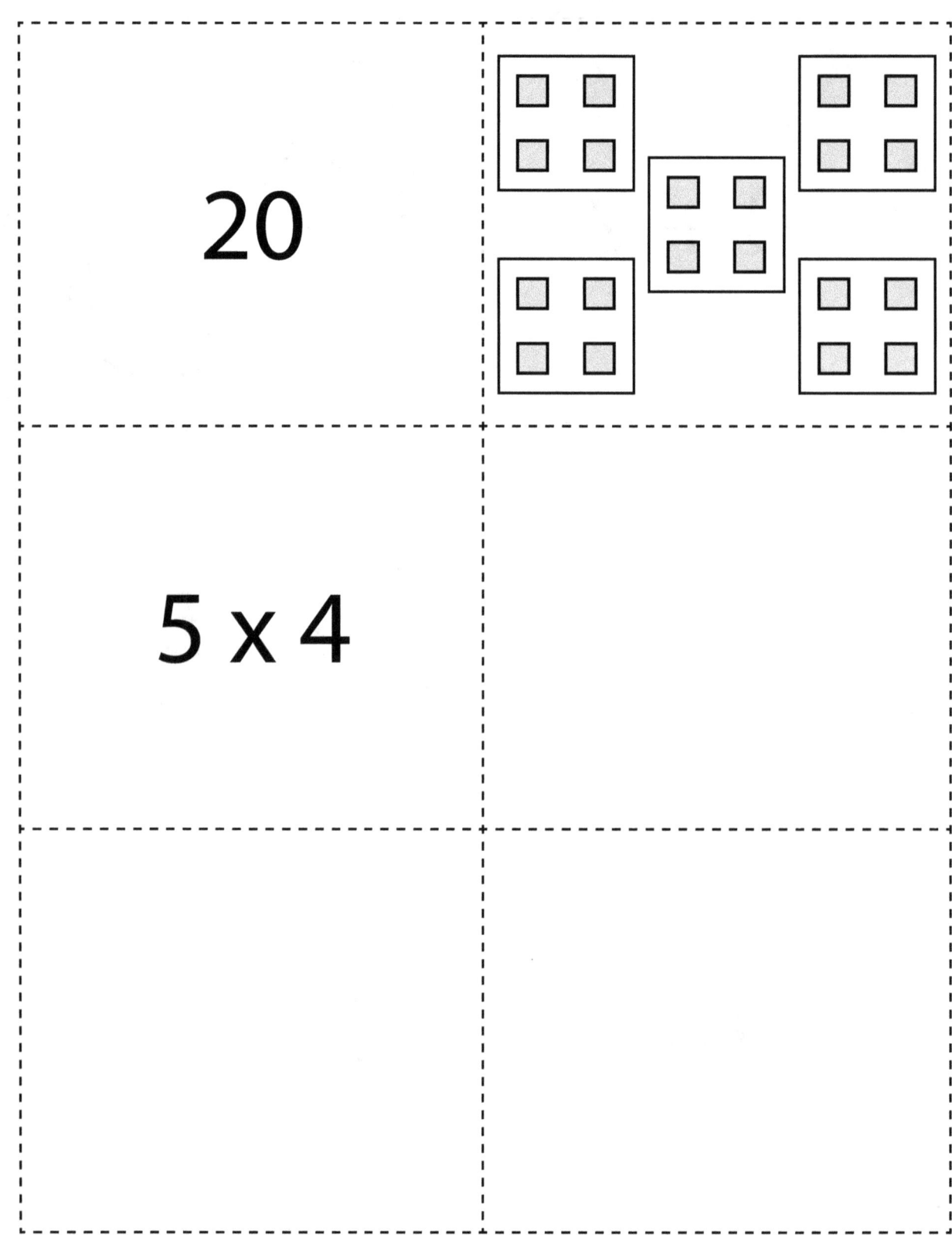

20

5 x 4

Equal Shares: Multiplication
RECORDING SHEET

Name: _____ Date: _____

Number Sentence	Picture	Solution

Shirley Disseler — Strategies and Activities for Common Core Math: Grades 3–5

EQUAL SHARES: DIVISION

GRADE 3

COMMON CORE STANDARD
3.OA.A.2

MATHEMATICAL PRACTICES
MP1
MP2

MATERIALS
• Equal Shares cards (remove multiplication sentence cards for this game)
• Equal Shares: Division recording sheet (one per student)

OVERVIEW

This game involves identifying groups of objects in a division problem. Common Core requires that students be able to **interpret groups or sets of objects as they refer to a division sentence**.

PROCEDURE

Game is played in pairs

1. Use the Equal Shares playing cards from the Equal Shares: Multiplication activity, but remove the multiplication sentence cards for this game.

2. Players shuffle the Equal Shares playing cards and deal six cards to each player. The remaining cards are placed face down as a draw pile.

3. Players review the cards they have been dealt to see if they can make a set of three, consisting of one number sentence, one picture card, and one solution card.

4. Players take turns, beginning with the youngest player.

5. If player 1 can make a set, he/she places the set down on the table and records the number sentence, picture, and solution on the recording sheet.

6. If no set can be made, the player discards one card to start a discard pile and draws a card from the draw pile.

7. Play continues until all cards are used.

8. Players count their sets. Winner has the greatest number of sets.

Equal Shares: Division
RECORDING SHEET

Name: _____ Date: _____

Number Sentence	Picture	Solution

RESOLVE TO SOLVE

GRADE 3

| COMMON CORE STANDARD |
| 3.OA.A.3 |

| MATHEMATICAL PRACTICES |
| MP1 |
| MP2 |
| MP4 |
| MP5 |
| MP6 |

MATERIALS
- Tiles or other manipulatives such as beans, beads, etc. (50 per pair of students)
- Resolve to Solve word problem cards (three cards per pair of students)
- Resolve to Solve recording sheet (one sheet for each problem assigned)

OVERVIEW

This activity promotes critical thinking in problem solving as students **model word problems and draw out solutions**. Students will also begin to think algebraically as they create equations for the models of each problem.

PROCEDURE
Activity is done in pairs

1 Teacher assigns three word problems to each pair of students, which allows for differentiation based on students' skills.

2 Students use tiles or other manipulatives to model the problems. Then students draw what they have modeled on the recording sheet and show the problem as an equation.

3 Students share their answers with the class.

Resolve to Solve
WORD PROBLEM CARDS

1
There are 36 desks in the classroom. If the teacher puts 6 desks in each row, how many rows will there be?

2
There are 24 students in the class. The teacher wants to put them into 6 groups. How many will be in each group?

3
A roll of nickels contains 40 coins. How many sets of 25 cents can you make from one roll of nickels?

4
There are 60 people at the meeting. If each row holds 15 people, how many rows are there?

5
To play a game, the teacher divides the students in the class into 3 lines with 7 in each line. How many students are in the class?

6
25 students are in each third grade class. There are 4 classes of third graders in the school. How many third grade students are in the school?

Resolve to Solve
WORD PROBLEM CARDS

7
80 people attended science night. If there were 8 stations for people to attend and each station had an equal number of visitors, how many people went to each station?

8
There are 90 second graders in the school. If each second grade class has 30 students, how many second grade classes are there?

9
John has 55 baseball cards. If he wants to share equally with 5 friends, how many cards will each friend get?

10
There are 40 books on the reading list for third grade students. If a student reads 2 books each week, how many weeks would it take for them to read all of the books?

11
One hundred people attended the recital. If they sat in rows of 20, how many rows were there in all?

12
Mary's mom is making cupcakes for the PTA meeting. If each person eats 2 and there are 35 people coming to the meeting, how many cupcakes should she make?

Resolve to Solve
RECORDING SHEET

Name: _____ Date: _____

PROBLEM

_____
_____
_____
_____
_____

MODEL OF THE PROBLEM

```
┌─────────────────────────────────────────────────────────┐
│                                                         │
│                                                         │
│                                                         │
│                                                         │
│                                                         │
│                                                         │
│                                                         │
└─────────────────────────────────────────────────────────┘
```

EQUATION

_____
_____
_____
_____
_____

MYSTERY NUMBERS

GRADE 3

COMMON CORE STANDARDS
3.OA.A.4
3.OA.B.6

MATHEMATICAL PRACTICES
MP1
MP2

MATERIALS
- Deck of playing cards
- Mystery Numbers boards 1, 2, 3, and 4 (one per student)
- Beans or items to cover squares on the number sentences (ten per student)
- Overhead projector or document camera

OVERVIEW

This is a whole class review activity. Students are seeking to **demonstrate understanding of unknown numbers in division and multiplication problems**.

PROCEDURE

1. Each student has one Mystery Numbers board.
2. Teacher uses overhead projector or document camera to display playing cards drawn.
3. Teacher reveals one playing card at a time (as if playing bingo).

   Example
   Teacher displays *6 of hearts*
   Students look on their boards for squares with:
   $18 \div \underline{} = 3$ *(red)*   or $3 \times 2 = \underline{}$ *(red)*

4. Students cover appropriate solutions on their boards.
5. When students cover three in a row (across, up and down, or diagonally), they call out, "Mystery Solved!"

Mystery Numbers Board 1

18 ÷ ___ = 3 (RED)	4 x ___ = 36 (BLACK)	45 ÷ 5 = ___ (RED)
63 ÷ ___ = 7 (BLACK)	___ x 7 = 56 (RED)	___ ÷ 2 = 5 (BLACK)
60 ÷ ___ = 6 (RED)	12 x ___ = 84 (BLACK)	100 ÷ 10 = ___ (RED)

Mystery Numbers Board 2

21 ÷ ___ = 3 (RED)	6 x ___ = 36 (BLACK)	50 ÷ 5 = ___ (RED)
63 ÷ ___ = 9 (BLACK)	___ x 8 = 56 (RED)	10 ÷ ___ = 5 (BLACK)
80 ÷ ___ = 8 (RED)	11 x ___ = 77 (BLACK)	100 ÷ ___ = 20 (RED)

Mystery Numbers Board 3

27 ÷ ___ = 3 (RED)	4 x ___ = 32 (BLACK)	40 ÷ 5 = ___ (RED)
56 ÷ ___ = 8 (BLACK)	___ x 7 = 49 (RED)	40 ÷ ___ = 5 (BLACK)
12 ÷ ___ = 3 (RED)	8 x ___ = 64 (BLACK)	90 ÷ ___ = 9 (RED)

Mystery Numbers Board 4

27 ÷ 9 = ____ (RED)	____ x 8 = 32 (BLACK)	40 ÷ ____ = 8 (RED)
56 ÷ 8 = ____ (BLACK)	5 x 7 = ____ (RED)	40 ÷ ____ = 5 (BLACK)
24 ÷ ____ = 3 (RED)	14 ÷ ____ = 7 (BLACK)	48 ÷ ____ = 6 (RED)

DISTRIBUTIVE PROPERTY MATCH-UP

GRADES 3, 5

COMMON CORE STANDARDS
3.OA.B.5
5.OA.A.1

MATHEMATICAL PRACTICES
MP1
MP2
MP7

MATERIALS
• Distributive Property Match-Up problems

OVERVIEW

In this whole-class activity, students learn to **recognize properties in numerical expressions**, with particular emphasis on the **distributive property**. Use it to pre-assess and post-assess what students know. This activity also helps students learn to solve problems and determine that both sides are equal to the same thing.

PROCEDURE

1. Divide students into two equal groups—one for problems using the distributive property and one for non-distributive problems. Line up each group on opposite sides of the room.

2. Give each student one problem.

3. When teacher says, "go," students move around the room and match up with the person holding the problem that yields the equivalent solution.

4. Once students think they have found the person with the matching problem, each pair of students works together, discussing why their problems match.

5. Teacher leads a whole-class discussion about why the problems are equivalent, using the distributive format and the non-distributive format.

NOTE

Create more problems, if necessary, to have one problem for each student (either non-distributive or distributive). The set in this book will cover eighteen students.

Distributive Property Match-Up Problems

Non-distributive	Distributive
5 x (6 + 9)	(5 x 6) + (5 x 9)
12 x (2 + 3)	(12 x 2) + (12 x 3)
24 x (4 + 6)	(24 x 4) + (24 x 6)
3 x (2 + 5)	(3 x 2) + (3 x 5)
6 x (5 + 9)	(6 x 5) + (6 x 9)
2 x (5 + 3)	(2 x 5) + (2 x 3)
4 x (24 + 6)	(4 x 24) + (4 x 6)
2 x (12 + 3)	(2 x 12) + (2 x 3)
9 x (5 + 6)	(9 x 5) + (9 x 6)

SPEED SETS

GRADE 3

COMMON CORE STANDARD
3.OA.B.5

MATHEMATICAL PRACTICES
MP1
MP2

MATERIALS
• Speed Sets playing cards (one set per group)
• Timer

OVERVIEW

This game provides practice with the concept of **multiplication and division as inverse operations that belong to the same fact family**.

PROCEDURE

Game is played in groups of 3

1 Players shuffle the Speed Sets playing cards and spread them out randomly face up on the desk.

2 Teacher sets the timer for one minute.

3 When the timer starts, all players look for cards that form a two-card set (multiplication sentence and division sentence in the same fact family) and take those cards off the board.

4 When the time ends, each player counts the number of sets he/she has picked up.

5 The winner is the player with the most sets.

DIFFERENTIATION

The game can also be played in the "Concentration" format; that is, the cards are spread out face down, and players take turns turning over two cards, looking for matches.

Speed Sets
PLAYING CARDS

7×3	$21 \div 3$
4×5	$20 \div 5$
4×6	$24 \div 6$

Speed Sets
PLAYING CARDS

8 x 7	56 ÷ 7
6 x 4	24 ÷ 4
8 x 3	24 ÷ 3

Speed Sets
PLAYING CARDS

9 x 4	36 ÷ 4
3 x 5	15 ÷ 5
9 x 8	72 ÷ 8

Speed Sets
PLAYING CARDS

7 x 8	56 ÷ 8
4 x 12	48 ÷ 12
12 x 5	60 ÷ 5

Speed Sets
PLAYING CARDS

9 x 7	63 ÷ 7
6 x 8	48 ÷ 8
3 x 9	27 ÷ 9

Speed Sets
PLAYING CARDS

8 x 6	48 ÷ 6
4 x 7	28 ÷ 7
6 x 5	30 ÷ 5

Speed Sets
PLAYING CARDS

5 x 2	10 ÷ 2
7 x 6	42 ÷ 6
6 x 3	18 ÷ 3

MODELING PROPERTIES

GRADES 3, 4

COMMON CORE STANDARDS 3.OA.B.6 4.NBT.B.5

MATHEMATICAL PRACTICES MP1 MP4 MP7

MATERIALS • Color tiles • Crayons • Modeling Properties problems • Modeling Properties recording sheet (one per student or pair) • Commutative and Associative Properties chart

OVERVIEW

This activity assesses student **understanding of property relationships** using the Associative and Commutative properties of addition and multiplication.

PROCEDURE

Activity is done individually or in pairs

1 Display one of the Modeling Properties problems.

2 Students use tiles to model the problem.

3 Students use crayons to record their answers on the recording sheet. Use the problems given and create more for additional challenge.

NOTE

Display or distribute copies of the Commutative and Associative Properties chart to show students how to model the problems.

Commutative and Associative Properties

COMMUTATIVE PROPERTY OF ADDITION

The sum of a set of numbers is not affected by the order of the numbers.

    4 + 2 = 2 + 4

○ ○ ●   ● ○ ○
○ ○ ● = ● ○ ○

COMMUTATIVE PROPERTY OF MULTIPLICATION

The product of a set of numbers is not affected by the order of the numbers.

    3 x 2 = 2 x 3

○ ○ ○   ○ ○ ○ = ○ ○   ○ ○   ○ ○

ASSOCIATIVE PROPERTY OF ADDITION

Changing the groupings of addends does not change the sum.

    4 + (6 + 2) = (4 + 6) + 2

○○○○ + (●●●●●● ●●) = (○○○○ ●●●●●●) + (●●)

ASSOCIATIVE PROPERTY OF MULTIPLICATION

Changing the groupings of factors does not change the product.

    (3 x 2) x 4 = 3 x (2 x 4)

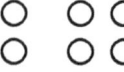

Modeling Properties
PROBLEMS

Associative Property of Addition:

problem 1: $3 + (2 + 6) = (3 + 6) + 2$
problem 2: $(2 + 4) + 3 = 2 + (3 + 4)$
problem 3: $4 + (2 + 5) = (4 + 2) + 5$

Commutative Property of Addition:

problem 1: $2 + 4 = 4 + 2$
problem 2: $7 + 3 = 3 + 7$
problem 3: $6 + 4 = 4 + 6$

Associative Property of Multiplication:

problem 1: $(3 \times 2) \times 4 = 3 \times (2 \times 4)$
problem 2: $7 \times (2 \times 3) = (7 \times 2) \times 3$
problem 3: $(4 \times 2) \times 4 = 4 \times (2 \times 4)$

Commutative Property of Multiplication:

problem 1: $4 \times 3 \times 2 = 3 \times 4 \times 2$
problem 2: $6 \times 4 = 4 \times 6$
problem 3: $2 \times 6 \times 5 = 5 \times 2 \times 6$

Modeling Properties
RECORDING SHEET

Name: _____ Date: _____

Problem	Property	Model/Picture

ORDER UP

GRADES 3, 5

COMMON CORE STANDARDS
- 3.OA.C.7
- 3.NBT.A.2
- 5.OA.A.1

MATHEMATICAL PRACTICES
- MP1
- MP5
- MP7

MATERIALS
- Deck of cards (one per group of students)
- Paper and pencil (one per student)
- Order Up recording sheet (one per student)
- Calculator (one per student)

OVERVIEW

This game helps students practice and review how to **use parentheses, brackets, and braces** in the evaluation of expressions. Students also **utilize the operational symbols** of mathematics.

PROCEDURE

Play in groups of 2 – 3

1. Students shuffle and deal five cards to each player. Place remainder of cards in a pile face down on the table.

2. Turn over the top card. This card is the target number for the round.

3. Using the numbers on the cards in their hands, each player writes an expression on his/her recording sheet to equal the target number. Players should use as many operational symbols as possible. Players may insert up to two sets of parentheses into the problem. Players should use as many cards as possible in each round. Each player must have five cards at all times.

4. If a player cannot make the problem work, he/she may discard one card from his/her hand and draw another card, losing one point. The player may then try once more to make an expression.

5. Once all players have presented their problems, each person checks with a calculator to make sure the problems are correct.

6. Points are tallied and recorded (see scoring).

7. Cards are returned to the deck, reshuffled and a new target number is drawn for round 2.

8. The player with the highest score after five rounds is the winner.

SCORING

- Five points are awarded for a correct answer.
- One additional point is awarded for the use of each different symbol, including parentheses.
- If player discards and draws another card, player loses one point.
- Points and problems are recorded on the students' recording sheets.

    Ace = 1   Jack = 11   Queen = 12   King = 13

DIFFERENTIATION

For additional difficulty and for those who are ready for integers, red cards can represent negative numbers and black cards represent positive numbers.

Order Up
RECORDING SHEET

Name: _____ Date: _____

Round	Problem and Solution	Total Points
1		
2		
3		
4		
5		
	Total:	

SCORING

- 5 points for correct answer
- 1 additional point for each different symbol, including parentheses.
- Player loses 1 point if he/she discards a card and draws another

        Ace = 1   Jack = 11   Queen = 12   King = 13

WORD PROBLEM FRENZY

GRADES 3, 4

COMMON CORE STANDARDS
3.OA.D.8
4.OA.A3

MATHEMATICAL PRACTICES
MP1
MP2
MP3

MATERIALS
• Word Problem Frenzy problems (one set per student or pair)
• Word Problem Frenzy equations (one set per student or pair)
• Word Problem Frenzy recording sheet (2 per student if all 10 problems are used)

OVERVIEW

This activity helps begin the process **of solving two-step word problems using all four operations**. Students should first be able to identify the correct operation for a given problem. This activity can also **serve as a formative assessment tool**.

This activity can be done in pairs to promote conversation, or individually as an assessment tool.

PROCEDURE
Activity is done in pairs or individually

1 Students match up the problems with the equations by cutting out the problems and the equations and pasting the matching ones on the recording sheet.

2 On the recording sheet, write a brief explanation of why the chosen equation is correct for the problem.

3 Students share their thought processes with other students.

Word Problem Frenzy
WORD PROBLEMS

1
If John brought 20 cupcakes to the party and Sue had already delivered 15 cupcakes, how many cupcakes were there in all?

2
On the last day of school, 24 students showed up out of 27. Eight of the students checked out early. How many students were left for the rest of the day?

3
A school bus made 5 stops on its route. At each stop it dropped off 6 students. How many students were dropped off during the route?

4
Clay played in 12 basketball games during the season. He scored 8 points per game. How many points did he score in total?

5
Brianna bought 6 music albums online. If each album had 12 songs on it, how many songs did she purchase?

6
James mows lawns in the summer to make some extra cash. He made $200 last summer. If he charges $20 per lawn, how many lawns did he mow?

7
If Martin has 63 books on his shelf to read and he reads 3 books a day, how many days will it take him to read all of his books?

8
Mary babysits to make money on the weekends. She charges $6 per hour. If she wants to make $60 this weekend, how many hours does she need to babysit?

9
At the fair, Ali bought 6 tickets for the Ferris wheel, 4 tickets for the scrambler, and 2 tickets for the dunking booth. How many tickets did she buy in all?

10
Three hundred people went to the concert at the park. If 120 of them left early, how many were left by the end of the concert?

Word Problem Frenzy
EQUATIONS

20 + 15	20 - 15	20 x 15
6 - 5	5 x 6	5 + 6
6 x 12	12 ÷ 6	12 - 6
63 ÷ 3	12 + 6	63 - 3
63 + 3	63 x 3	6 + 4 - 2
24 – 8	24 + 8	6 + 4 + 2

Word Problem Frenzy
EQUATIONS

12 x 8	12 + 8	12 - 8
200 ÷ 20	200 - 20	200 + 20
200 x 20	60 ÷ 6	60 - 6
60 x 6	60 + 6	300 ÷ 120
300 + 120	300 - 120	

Word Problem Frenzy
RECORDING SHEET

Name: _____ Date: _____

Paste a word problem in the left column. Paste the matching equation in the center column. In the right column, explain why the equation matches the word problem.

Word Problem	Equation	Explanation

ANSWER SHEET

Word Problem Frenzy
RECORDING SHEET

Name: _____ Date: _____

Paste a word problem in the left column. Paste the matching equation in the center column. In the right column, explain why the equation matches the word problem.

Word Problem	Equation	Explanation
1 If John brought 20 cupcakes to the party and Sue had already delivered 15 cupcakes, how many cupcakes were there in all?	20 + 15	John added 20 more cupcakes to the 15 Sue had already brought to the party, so you join the two numbers together to get the total.
2 On the last day of school, 24 students showed up out of 27. Eight of the students checked out early. How many students were left for the rest of the day?	24 - 8	Only 24 students were in school that day, and then 8 students left before the end of the day, so you subtract 8 from 24 to find out how many were left at the end of the day.
3 A school bus made 5 stops on its route. At each stop it dropped off 6 students. How many students were dropped off during the route?	5 x 6	The bus stopped 6 times, and every time it stopped, 5 students got out, so you multiply the numbers to find out how many students were dropped off altogether.
4 Clay played in 12 basketball games during the season. He scored 8 points per game. How many points did he score in total?	12 x 8	He played in 12 basketball games and scored 8 points every game he played in, so you multiply to find how many points he scored altogether.

ANSWER SHEET

Word Problem Frenzy
RECORDING SHEET

Name: _____ Date: _____

Paste a word problem in the left column. Paste the matching equation in the center column. In the right column, explain why the equation matches the word problem.

Word Problem	Equation	Explanation
5 Brianna bought 6 music albums online. If each album had 12 songs on it, how many songs did she purchase?	6 x 12	Every album she bought had 12 songs, and she bought 6 albums altogether, so you multiply the numbers to find how many songs there were in all the albums.
6 James mows lawns in the summer to make some extra cash. He made $200 last summer. If he charges $20 per lawn, how many lawns did he mow?	200 ÷ 20	He earns $20 each time he mows, and he earned $200 altogether, so you decompose 200 into sets of 20 and count the sets to find how many times he mowed.
7 If Martin has 63 books on his shelf to read and he reads 3 books a day, how many days will it take him to read all of his books?	63 ÷ 3	He reads 3 books each day and has 63 books altogether, so you divide 63 into sets of 3 and count the sets to find how many days it will take to read all 63 books.
8 Mary babysits to make money on the weekends. She charges $6 per hour. If she wants to make $60 this weekend, how many hours does she need to babysit?	60 ÷ 6	She wants to make $60 in total and she earns $6 an hour, so you divide 60 into sets of 6 and count the sets to find how many hours she babysits to earn that much.

Shirley Disseler   Strategies and Activities for Common Core Math: Grades 3–5   Part 1

ANSWER SHEET

Word Problem Frenzy
RECORDING SHEET

Name: _____ Date: _____

Paste a word problem in the left column. Paste the matching equation in the center column. In the right column, explain why the equation matches the word problem.

Word Problem	Equation	Explanation
9 At the fair, Ali bought 6 tickets for the Ferris wheel, 4 tickets for the scrambler, and 2 tickets for the dunking booth. How many tickets did she buy in all?	$6 + 4 + 2$	You add all the tickets together to find how many she bought in total.
10 Three hundred people went to the concert at the park. If 120 of them left early, how many were left by the end of the concert?	$300 - 120$	Since 300 people came to the concert but 120 left it early, you subtract 120 from 300 to find how many people were still at the concert at the end.

64 Strategies and Activities for Common Core Math: Grades 3–5   Part 1   Shirley Disseler

VENN NUMBER THEORY

GRADES 3, 4, 5

COMMON CORE STANDARDS 3.OA.D.9 4.OA.B.4 5.OA.B.3

MATHEMATICAL PRACTICES MP1 MP2 MP3 MP6

MATERIALS • Venn diagram • Number tiles

OVERVIEW

This whole-class activity helps students learn about **patterns in numbers**, using such terms as *prime, composite, sum of the digits, differences,* and *factors*.

The general procedure is for the teacher to place numbers on the Venn diagram and have students discuss the relationship of those numbers. Start by demonstrating this activity with the whole class. Use examples A, B, or C, or demonstrate another number relationship. Then have the students work in groups or individually, using the Venn diagram and number tiles.

PROCEDURE

Example A
Prime and composite or odd and even numbers - grades 3 and 4

1. On the left side of the Venn diagram place: 3, 7, 11, 17 (prime or odd numbers)
2. On the right side of the Venn diagram place: 4, 6, 8, 10 (composite or even numbers)
3. Ask students where 2 might be placed and why. Correct answer would be "in the middle because it is even and prime."

Example B
Divisibility rules - grade 5

1. On the left side of the Venn diagram place: 64, 46, 10, 16, 44, 28 (divisible by 2)
2. On the right side of the Venn diagram place: 99, 21, 87, 69 (divisible by 3)
3. In the center of the Venn diagram place: 12, 48, 66 (divisible by both 2 and 3)
4. As you place numbers, ask students if they can determine the concept being explored.
5. Discuss divisibility rules.

Example C
Ordered pairs - grade 5

1. On the left side of the Venn diagram place: 0, 3, 6
2. On the right side of the Venn diagram place: 3, 6, 9
3. Ask students to identify the rule that creates a relationship

between each side. (Correct answer would be "the rule is x + 3, given that pair one is (0,3), pair two is (3,6), and pair three is (6,9).")

DIFFERENTIATION

Have students create their own number relationships for other students to solve.

Venn Diagram

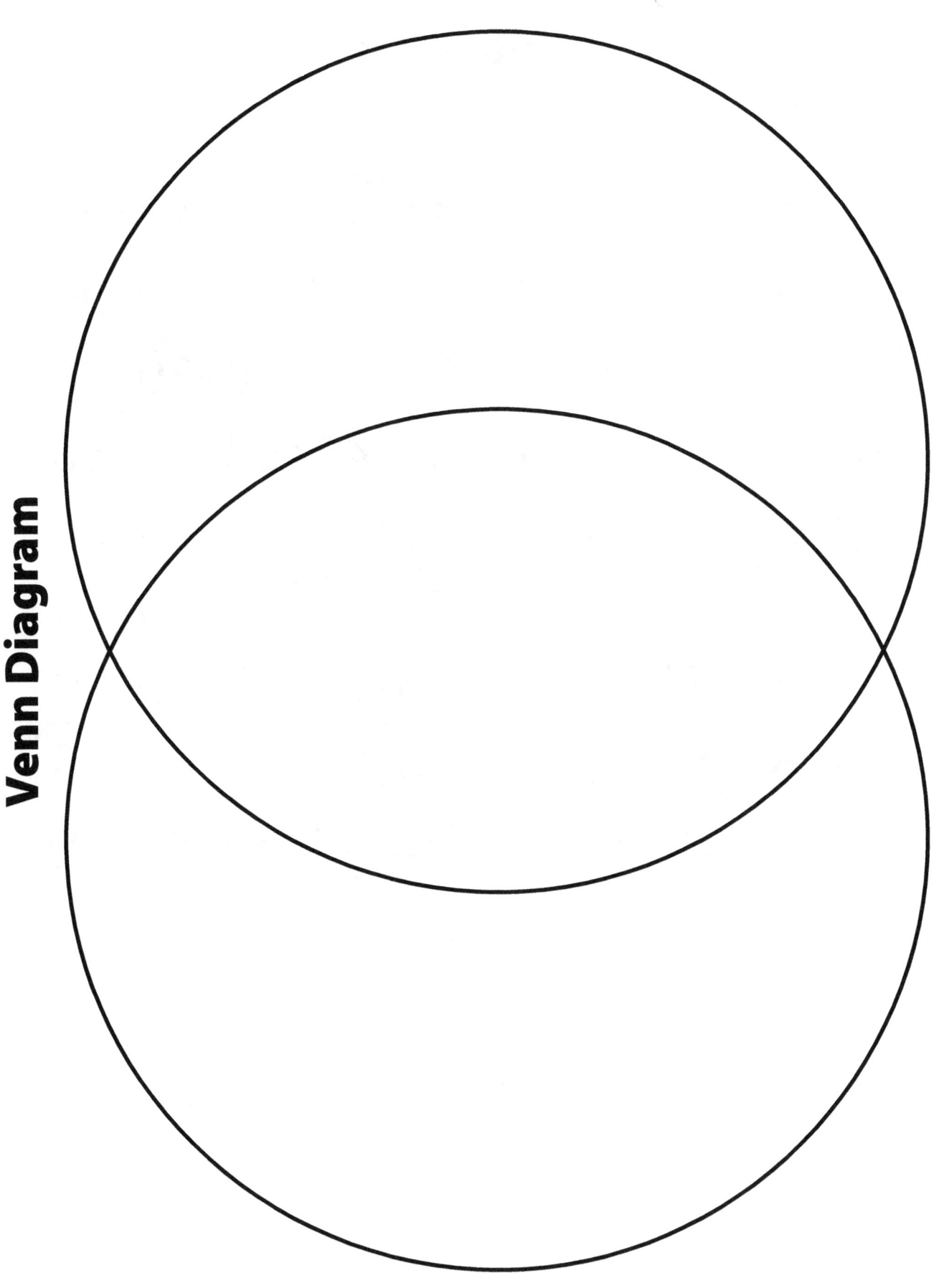

Number Tiles

1	2	3	4	5	6	7	8	9	10
11	12	13	14	15	16	17	18	18	20
21	22	23	24	25	26	27	28	29	30
31	32	33	34	35	36	37	38	39	40
41	42	43	44	45	46	47	48	49	50
51	52	53	54	55	56	57	58	59	60
61	62	63	64	65	66	67	68	69	70
71	72	73	74	75	76	77	78	79	80
81	82	83	84	85	86	87	88	89	90
91	92	93	94	95	96	97	98	99	100

VISUAL PATTERNING

GRADE 3

COMMON CORE STANDARD
3.OA.D.9

MATHEMATICAL PRACTICES
MP2
MP4
MP5
MP7

ACTIVITY 1 MATERIALS
• Addition table (one per student)
• Highlighters

ACTIVITY 2 MATERIALS
• Multiplication table (one per student)
• Highlighters

OVERVIEW

In this whole-class activity, students learn about how **pattern relationships** exist within the number system. This will lead to a **deeper understanding of the commutative, associative, and distributive properties**.

ACTIVITY 1 – PROCEDURE

1 Students highlight numbers on the addition table to show even numbers as two equal addends.

$6 = 3 + 3$
$18 = 9 + 9$

2 Make sure students notice that all sums in an addition table increase by the same amount (by 1).

ACTIVITY 2 (ADVANCED) – PROCEDURE

1 Students highlight numbers on the multiplication table to show that multiples of even numbers are always equivalent to even numbers.

$10 = 20, 30, 40$
$8 = 16, 24, 32$

2 Students highlight the *8* column and row. Make sure they notice that either way the product is the same (commutative property).

3 Students highlight the *1* and *2* rows on the table. Think of these as fractions: ½, 2/4, 3/6, etc.
  a. Ask students what they notice about these numbers. (Correct answer: They are all equivalent fractions.)
  b. Students test two other rows to see if the pattern continues. (Correct answer: It does.)

4 Students draw a rectangle around a set of at least four numbers anywhere on the board. Ask students if they can find a pattern in the four numbers. (Correct answer: The numbers in the four corners of the rectangle the students have drawn are proportional.)

Addition Table

+	1	2	3	4	5	6	7	8	9	10
1	2	3	4	5	6	7	8	9	10	11
2	3	4	5	6	7	8	9	10	11	12
3	4	5	6	7	8	9	10	11	12	13
4	5	6	7	8	9	10	11	12	13	14
5	6	7	8	9	10	11	12	13	14	15
6	7	8	9	10	11	12	13	14	15	16
7	8	9	10	11	12	13	14	15	16	17
8	9	10	11	12	13	14	15	16	17	18
9	10	11	12	13	14	15	16	17	18	19
10	11	12	13	14	15	16	17	18	19	20

Multiplication Table

x	1	2	3	4	5	6	7	8	9	10	11	12
1	1	2	3	4	5	6	7	8	9	10	11	12
2	2	4	6	8	10	12	14	16	18	20	22	24
3	3	6	9	12	15	18	21	24	27	30	33	36
4	4	8	12	16	20	24	28	32	36	40	44	48
5	5	10	15	20	25	30	35	40	45	50	55	60
6	6	12	18	24	30	36	42	48	54	60	66	72
7	7	14	21	28	35	42	49	56	63	70	77	84
8	8	16	24	32	40	48	56	64	72	80	88	96
9	9	18	27	36	45	54	63	72	81	90	99	108
10	10	20	30	40	50	60	70	80	90	100	110	120
11	11	22	33	44	55	66	77	88	99	110	121	132
12	12	24	36	48	60	72	84	96	108	120	132	144

COMPARATIVE EQUATIONS

GRADE 4

COMMON CORE STANDARDS
4.OA.A.1
4.OA.A.2

MATHEMATICAL PRACTICES
MP1
MP2

MATERIALS
- Comparative Equations picture cards (one set per pair)
- Comparative Equations word problem cards (one set per pair)
- Comparative Equations game board (one per pair)
- Comparative Equations recording sheet (one per student)

OVERVIEW
Several Common Core Standards reference multiplicative comparisons. This activity **uses drawings and equations to solve multiplication and division word problems**.

PROCEDURE
Activity is done in pairs

1. Shuffle the word problem cards and the picture cards separately.
2. Place each stack of cards upside down in its appropriate box on the game board.
3. Each student draws two cards from each stack.
4. Student 1 begins the round by looking at the cards in his/her hand for a match of a word problem and a picture problem.
5. If the student has a matching set, he/she lays down the set and draws one more card from each stack.
6. If no match is made, the student discards one card and draws one from the stack of that type of card to maintain four cards in the hand.
7. When all the cards in the two stacks have been drawn, students explain their card matches to each other.
8. On the recording sheet, students write explanations for the matches they have made, as well as their equations and solutions.

Comparative Equations
PICTURE CARDS

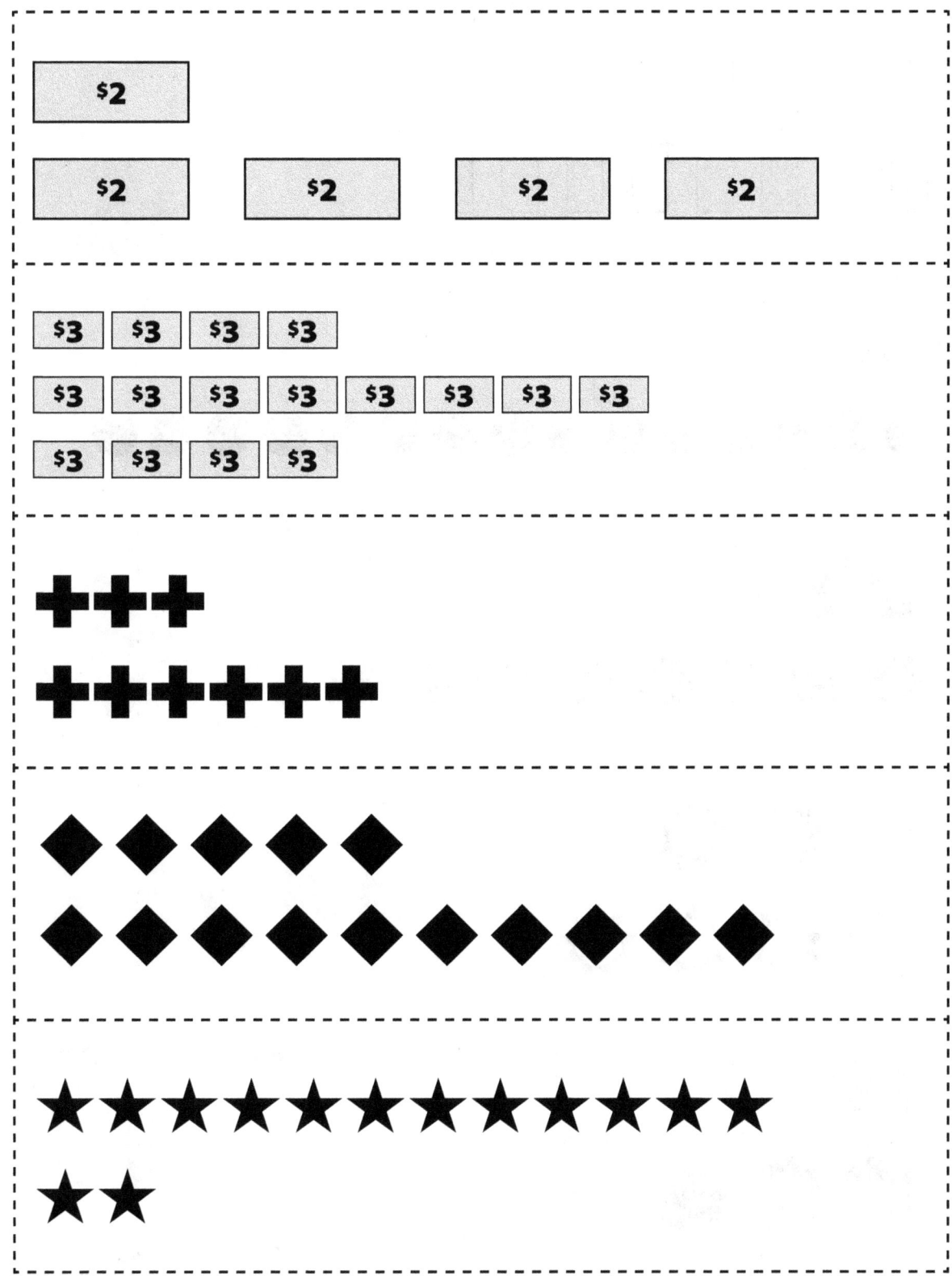

Shirley Disseler — Strategies and Activities for Common Core Math: Grades 3–5 — Part 1

Comparative Equations
PICTURE CARDS

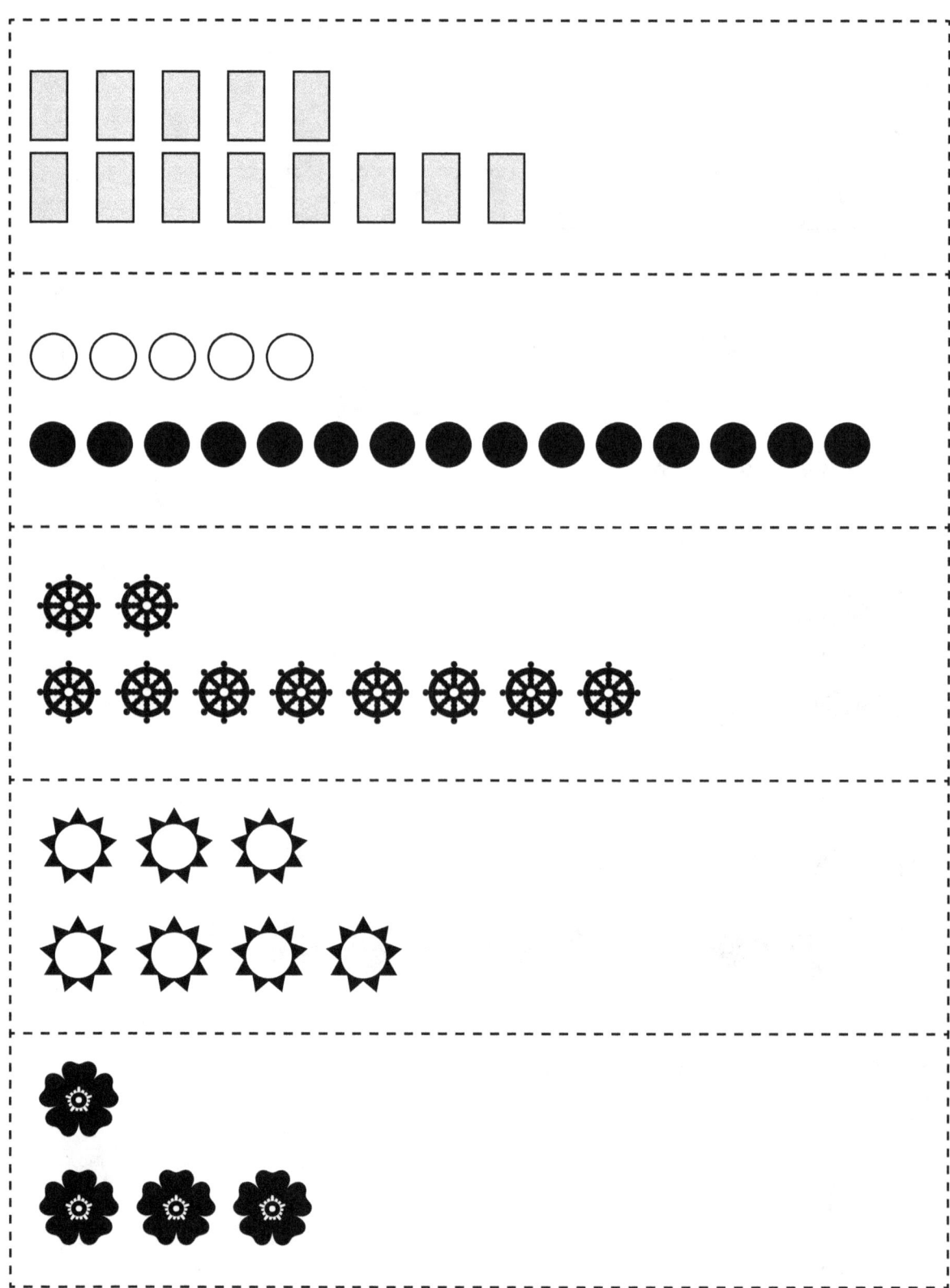

74 Strategies and Activities for Common Core Math: Grades 3–5   Part 1   Shirley Disseler

Comparative Equations
PICTURE CARDS

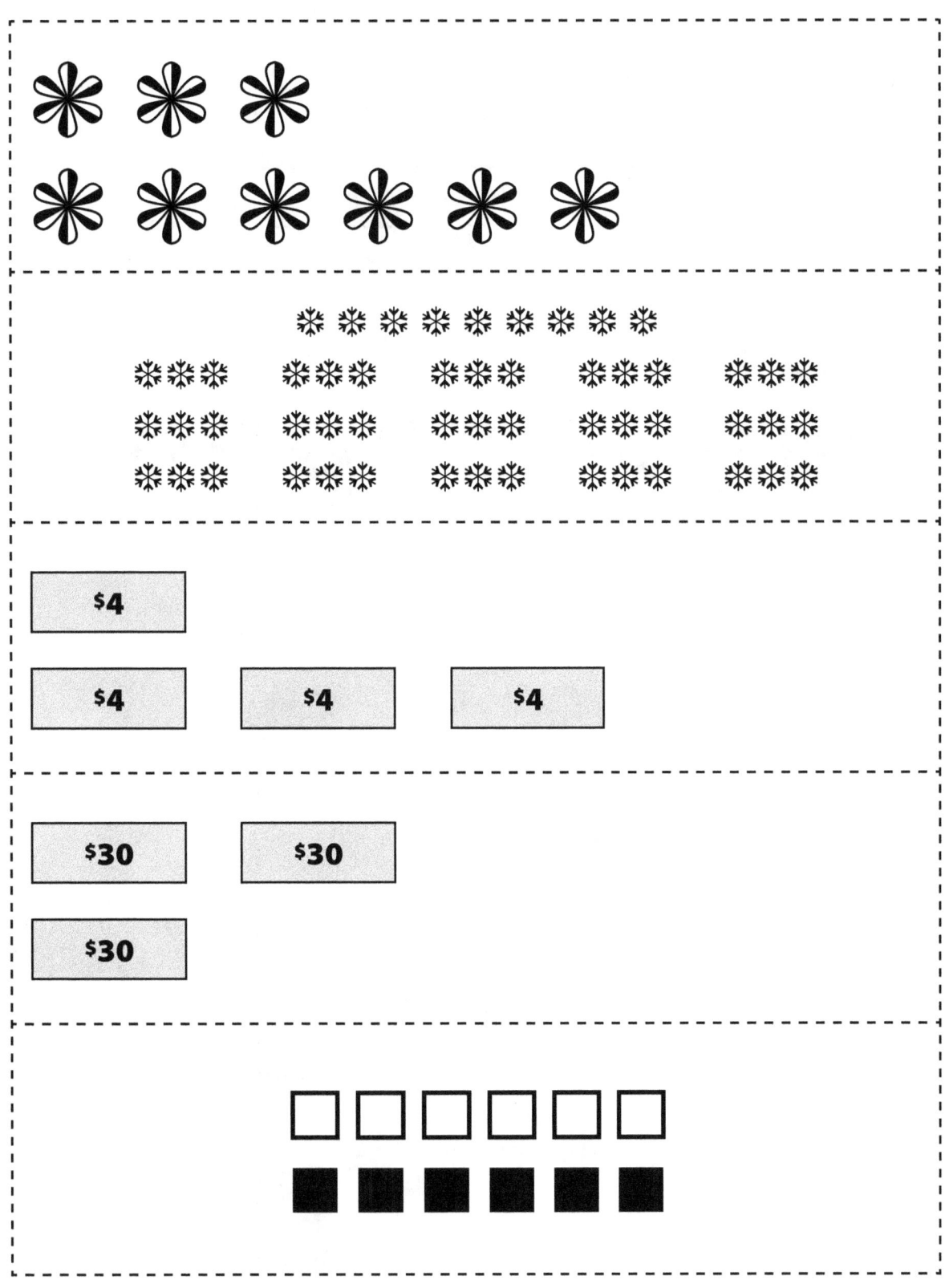

Shirley Disseler · Strategies and Activities for Common Core Math: Grades 3–5 · Part 1 · 75

Comparative Equations
PICTURE CARDS

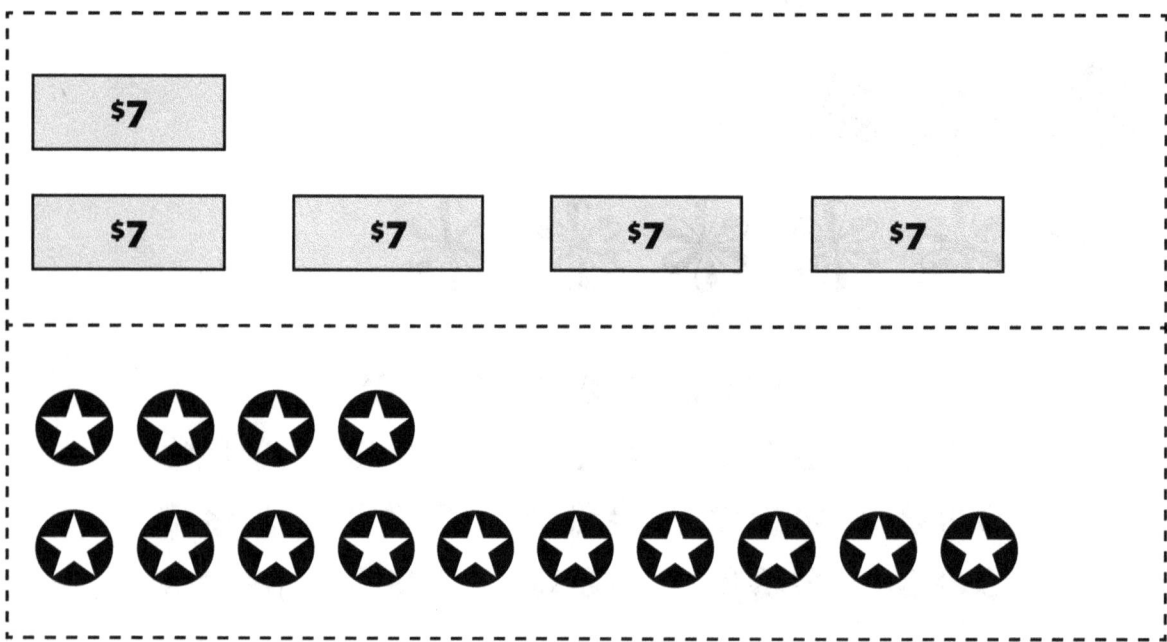

Comparative Equations
WORD PROBLEMS

One candy bar costs $2. How much would four times that amount cost?

A baseball card cost $3. John has 4 cards. Mark has three times as many cards as John has. How much does Mark's collection cost?

Sue is 3 years old and her sister is twice her age. How old is her sister?

Jack is one half his brother's age. His brother is 10 years old. How old is Jack?

Ali bought a dozen eggs. She used 1/6 of them to bake a cake. How many did she use?

Comparative Equations
WORD PROBLEMS

Tori sold three more boxes of popcorn than Emily. If Emily sold 5 boxes, how many did Tori sell?

The basket of apples at the store contained 20 apples. If 5 of them were bad, how many were not bad?

If every rabbit at the pet store has 4 babies, how many baby rabbits will there be if there are 2 rabbits in the store that are having babies?

Each hat is worth $10. Mary's hat collection is worth $40. Jonna has 1 hat fewer than Mary. How many hats does Jonna have?

One hamburger has 250 calories. If Liz eats three times as much as Ellie, and Ellie has one burger, how many calories did Liz eat?

Comparative Equations
WORD PROBLEMS

Callie had half as many ice cream cones as Anna. If Callie had 3 ice cream cones, how many did Anna have?

John's mom is five times older than John. If John is 9, how old is his mom?

Leslie spent $4 on snacks for the sleepover. Olivia spent three times more than Leslie. How much did Olivia spend?

Marty spent twice as much on his skateboard as Alan. If Marty spent $60, how much did Alan spend?

Keri bought 2 books every month for 6 months. If she read half of the books she bought, how many did she read?

Comparative Equations
WORD PROBLEMS

One can of paint cost $7. If it will take 4 cans to paint the office, how much will it cost in all?

One table in the restaurant seats 4 people. If two and half tables are needed for the dinner, how many people will attend?

Comparative Equations
GAME BOARD

Word Problem Cards

Picture Cards

Shirley Disseler

Comparative Equations
RECORDING SHEET

Name: _____ Date: _____

Word problem explanation	Equation/Solution

PRIME CIRCLES

GRADE 4

COMMON CORE STANDARD   **4.OA.B.4**

MATHEMATICAL PRACTICES   MP1   MP2   MP3

MATERIALS   • Prime Circles game board (one per pair of students)   • colored pencils (one per student)   • number cube or die (one per pair of students)   • game token (one per pair of students)   • optional: calculator (one per pair of students)   • Divisibility Rules chart

OVERVIEW

This game reinforces or introduces the concept **of prime and composite numbers**. It also helps to **review divisibility rules** and enhance the understanding of how the rules apply to numeration that includes prime numbers and composites.

PROCEDURE

Game is played in pairs

1 Each player chooses a colored pencil.

2 Player 1 chooses a number on the game board and places a token on it.

3 Player 1 rolls the number cube and decides if his/her number is divisible by that number.

4 If player 2 agrees with player 1, then player 1 shades in the number and removes the token. The calculator can be used to check the answer, if needed.

5 Player 2 takes a turn, proceeding with steps 2 - 5.

6 When a player captures three numbers in a row, he/she circles the three numbers. Numbers may not overlap or be used more than once.

7 Winner is the first player who circles five groups of three numbers.

NOTE

A chart of divisibility rules is included. Teachers may want to display this information as they teach the concepts or print copies for each student to reinforce the rules.

Shirley Disseler — Strategies and Activities for Common Core Math: Grades 3–5

Prime Circles
GAME BOARD

15	21	102	372	120	132	57	4
75	105	152	505	85	135	180	11
84	19	183	157	120	130	132	20
18	136	162	46	93	380	25	23
87	30	192	101	333	185	196	39
58	60	55	550	237	156	115	60
129	195	172	190	180	8	30	128
168	33	388	267	266	54	117	168
51	384	81	580	171	65	171	376
35	140	78	190	177	40	108	39
249	42	70	36	55	164	124	27

Divisibility Rules

PRIME NUMBER: Only divisible by 1 and itself

Examples – 1, 2, 3, 5, 7, 11

Only even prime number: 2

COMPOSITE NUMBER: Divisible by more than two factors

Examples – 4, 6, 8, 9, 10

NUMBERS DIVISIBLE BY:

2   End in an even number

3   Sum of the digits is divisible by 3
    Example
    15: 1 + 5 = 6 (15 and 6 are both divisible by 3)

4   Last two digits are divisible by 4
    Example
    384: 8 and 4 are each divisible by 4 (384 is divisible by 4)

5   End in 0 or 5

6   Is divisible by both 2 and 3

7   If you double the digit in the ones, then take that numbr and subtract the digit in the tens, the answer is 0 or divisible by 7
    Example
    56: 6 x 2 = 12 and 12 - 5 = 7 (56 is divisible by 7)

8   Last three digits are divisible by 8
    Example
    1248: 24 is divisible by 8 and 8 is divisible by 8 (1248 is divisible by 8)

9   Sum of the digits is divisible by 9
    Example
    54: 5 + 4 = 9 (54 is divisible by 9)

10 End in 0

PRIME OR COMPOSITE?

GRADE 4

COMMON CORE STANDARD
4.OA.B.4

MATHEMATICAL PRACTICES
MP1
MP2
MP3

MATERIALS
- Hundreds board (one per student)
- Snap cubes (15 per student)
- Prime or Composite? recording sheet (one per student)

OVERVIEW

This whole-class activity is a **review of prime and composite numbers**, and gives a physical demonstration of factorization. This hands-on manipulative strategy to teach prime factorization gives teachers the opportunity to **explore what students know about** numbers through vocabulary and questioning about **prime, composite, divisible, rules of divisibility, even, and odd.**

PROCEDURE

Use this activity first in a whole class setting. Once the students understand the activity, have them work in pairs or individually as you provide numbers. Walk around as the students do the activity to note who is having success and who needs more instruction. See example using the number *72*.

1. Each student places a snap cube in one hand.

2. Teacher calls out a composite number and students place the cube on the number on the hundreds board without removing the hand. Teacher asks, "Is this number prime or composite?" Students answer.

3. Teacher asks, "What two numbers can we multiply together to get this number?" Students answer.

4. Students slide the cube to one of the two numbers. Teacher repeats the "prime or composite?" question for each number.

5. Students place a cube on the prime number that makes up one of the two factors.

6. Repeat the process with the second number, as many times as necessary. This process will create the stack that shows the prime factorization.

7. Students write the number of cubes stacked on each prime number in the corresponding column of the Prime Factorization recording sheet and write the final prime factorization in the last column.

Prime or Composite?
EXAMPLE: 72

Teacher asks, "Is this number prime or composite?" If it is composite, teacher asks, "What two numbers can we multiply together to get this number?"

Students might say 2 x 36 = 72. On the hundreds board, students place one cube on 2, which is prime. Students place another cube on 36, which is composite.

Teacher asks, "Is this number (36) prime or composite?" If it is composite, teacher asks, "What two numbers can we multiply together to get this number?"

Students might say 6 x 6 = 36. Students move the cube from 36 to 6.

Teacher asks, "Is this number (6) prime or composite?" If it is composite, teacher asks, "What two numbers can we multiply together to get this number?"

Students might say 2 x 3 = 6. Students move the cube on 6 to 2, which is prime, creating a stack of two cubes on 2. Students place another cube on 3, which is prime.

Then teacher moves to the other 6. Students place another cube on 6.

Teacher asks, "Is this number (6) prime or composite?" If it is composite, teacher asks, "What two numbers can we multiply together to get this number?"

Students might say 2 x 3 = 6. Students move the cube on 6 to 2, which is prime, making a stack of three cubes on 2. Students place another cube on 3, which is prime, making a stack of two cubes on 3.

These stacks of cubes represent $2^3 \times 3^2 = 72$.

Students record their work on the Prime or Composite? recording sheet.

Prime or Composite?
RECORDING SHEET

Name: _____ Date: _____

Record the number for factorization in far left column. Write the number of cubes stacked on each number in the corresponding column. Write the final prime factorization in the far right column.

Number	2	3	5	7	11	Prime Factorization

Hundreds Board

1	2	3	4	5	6	7	8	9	10
11	12	13	14	15	16	17	18	18	20
21	22	23	24	25	26	27	28	29	30
31	32	33	34	35	36	37	38	39	40
41	42	43	44	45	46	47	48	49	50
51	52	53	54	55	56	57	58	59	60
61	62	63	64	65	66	67	68	69	70
71	72	73	74	75	76	77	78	79	80
81	82	83	84	85	86	87	88	89	90
91	92	93	94	95	96	97	98	99	100

HIT THE TARGET

GRADE 4

COMMON CORE STANDARD
4.OA.B.4

MATHEMATICAL PRACTICES
MP1
MP2
MP3

MATERIALS
- Hit the Target game board (one per pair of students)
- Hit the Target recording sheet (one per student)
- number cubes or dice (four per pair of students)
- tokens or small markers in two different colors (ten tokens for each player)

OVERVIEW

This game provides practice with **prime number identification, the creation of expressions, finding solutions to expressions, and analyzing solutions** to determine if they meet the criteria as a prime number.

PROCEDURE
Game is played in pairs

1 Students mark an "X" on all prime numbers on the game board.

2 Player 1 rolls four number cubes and writes the numbers on the recording sheet. Player 1 creates an expression equivalent to a prime number and writes the expression and its solution on the recording sheet.

   Example
   2, 3, 4, and 5 are rolled. Player could create the expression:
   $(5 \times 2) - (3 + 4) = 10 - 7 = 3$

3 Player 1 covers that prime number (*3* in the example above) with his/her token. If the player cannot create an expression, the player's turn is over.

4 Player 2 takes a turn.

5 The player with the most tokens on the board at the end of the time allotted is the winner.

Hit the Target
GAME BOARD

1	2	3	4	5	6	7	8	9	10
11	12	13	14	15	16	17	18	18	20
21	22	23	24	25	26	27	28	29	30
31	32	33	34	35	36	37	38	39	40
41	42	43	44	45	46	47	48	49	50
51	52	53	54	55	56	57	58	59	60
61	62	63	64	65	66	67	68	69	70
71	72	73	74	75	76	77	78	79	80
81	82	83	84	85	86	87	88	89	90
91	92	93	94	95	96	97	98	99	100

Hit the Target
RECORDING SHEET

Name: _____ Date: _____

Cube 1	Cube 2	Cube 3	Cube 4	Expression	Solution

THE NUMBER IS IN THE WORD

GRADE 5

COMMON CORE STANDARD
5.OA.A.2

MATHEMATICAL PRACTICES
MP1
MP2
MP6
MP7

MATERIALS

- The Number Is in the Word word strips (one set per group or class)
- The Number Is in the Word number strips (one set per group or class)
- The Number Is in the Word Recording Sheet (one per group for group activity)

OVERVIEW

This activity can be used as a whole class pre-assessment of student word-to-number expressions transfer, or it can be used in small group settings as an activity to build upon word-to-expression transfer. Students are working toward **writing expressions that express proper calculation order using words.** This is a difficult task, because students have often learned to work left to right in calculating, which is not always the correct order.

WHOLE-CLASS PROCEDURE

1. Split the class in half. One half goes to one side of the room and the other half to the other side of the room.
2. Give one half the word strips and the other half the number strips.
3. Students walk around without talking and find their matching partner.
4. Student stand in pairs when they have found their matching expressions.
5. When all students have found their matching expressions, each pair of students explains why they match.

SMALL-GROUP PROCEDURE

1. Divide class into groups of four students each. Give each group one set of both word strips and number strips.
2. Set a time limit for students to find matching sets. (Recommended time limit: 10 minutes)
3. When time is up, students remove unmatched word and number strips.
4. Students record the matching sets on the recording sheet.

NOTE

When creating the sets of word strips and number strips, copy the word strips on one color of paper or card stock and number strips on a different color.

The Number Is in the Word
WORD STRIPS

Two more than four times a number

The difference between seven and two times a number

The product of seven and a number

The quotient of six times a number and two

The sum of two times a number and one hundred

The quotient of twenty-four and a number

The difference between sixteen and a number

Two less than four times a number

The product of six and a number

The sum of y and fourteen

The quotient of seven and a number

The sum of two times a number and five

The quotient of three times a number and three

A number less than seventeen

The quotient of the sum of three and five divided by two

The sum of five times a number and six

The quotient of a number and twenty-four

A number less than sixteen

The Number Is in the Word
NUMBER STRIPS

$4x + 2$
$7 - 2x$
$7a$
$6t \div 2$
$100 + 2s$
$24 \div s$
$16 - n$
$4a - 2$
$6p$
$14 + y$
$7 \div r$
$5 + 2x$
$3t \div 3$
$17 - p$
$\dfrac{3 + 5}{2}$
$6 + 5x$
$y/24$
$16 - n$

The Number Is in the Word
RECORDING SHEET

Name: _____ Date: _____

Words	Expressions

NUMBER & OPERATIONS IN BASE TEN

This chapter has activities to help teach Common Core Standards in the domain of Number & Operations in Base Ten. This domain is closely linked to Operations & Algebraic Thinking. Several of the activities in the previous chapter are also aligned with standards in the Number & Operations in Base Ten domain.

The focus of this domain is on place value, properties of operations, and performing all four operations in mathematics using both whole numbers and decimals. Students' ability to show and prove answers is paramount. Solutions are only important if students can identify what they truly mean in terms of the math.

The teaching goals for this domain should center on student understanding of the "why" of numeration and operations performed with numbers. The procedural knowledge of this domain should only become the focus after students have explored their own methods for solving problems using inventive strategies. Once students have explored numbers in base ten, teachers can begin the process of computation. Several processes should be explained—no one process should be considered the only way to solve a problem.

One of the key strategies necessary for students to obtain success with Number & Operations in Base Ten is modeling. Teachers should model problems and the strategies for solving them, and then teach students to do the same.

Students need a firm foundation in place value to understand what base ten really means. If students understand the powers of ten in terms of place value, decimals will prove much easier to understand. This chart can help demonstrate how decimal numeration fits into base ten as students move into decimals from whole numbers:

Hundreds	Tens	Ones	.	Tenths	Hundredths	Thousandths
10^2	10^1	10^0	.	10^{-1}	10^{-2}	10^{-3}
		3	.	4		
	2	3	.	4	5	
1	2	3	.	4	5	6

Students must be obtain the connection between decimals and fractions simultaneously to formulate a true relationship between the two. One of the purposes of this Common Core domain is to help students see that the same number can be written in several ways. Students should convert fractions to decimals and decimals to fractions and see the numbers as equivalent amounts. Working with decimals should always be done in conjunction with whole numbers and fractions to help students link them.

This chapter contains many activities to provide students with opportunities to use all types of numbers in base ten. The games and activities in this chapter utilize such materials as dominoes, playing cards, and base ten blocks. For teachers who are unfamiliar with the use of base ten blocks, an example showing how to use them to model is included.

WHO'S IN WHOSE PLACE?

GRADES 3, 4, 5

COMMON CORE STANDARDS
3.NBT.A.1
4.NBT.A.1
4.NBT.A.2
4.NBT.A.3
5.NBT.A.2
5.NBT.A.3
5.NBT.A.4

MATHEMATICAL PRACTICES
MP5
MP6
MP7

MATERIALS
- Set of dominoes (one per group of students)
- Who's in Whose Place? recording sheet (one per group of students)
- Who's in Whose Place? place value mats: whole numbers and decimals (one per group of students)

OVERVIEW
This group activity gives students practice with **place value identification using multi-digit whole numbers.**

PROCEDURE: WHOLE NUMBERS
Activity is done in groups of 2 – 4

1. Students turn all the dominoes upside down and shuffle them.
2. Each student chooses enough dominoes to complete the number of places needed on the whole number mat (three dominoes for 6 places, four dominoes for 8 places).
3. Students arrange their dominoes to form a number with each side of the domino representing a number corresponding to the number of dots on each side.
4. Students write the number they have arranged on the recording sheet.
5. Students compare their numbers with the rest of the group.
6. Student with the highest number is the winner of the round.
7. All dominoes are placed back in the pile and shuffled again for a new round.

PROCEDURE: DECIMALS
Activity is done in groups of 2 – 4

1. Students turn all the dominoes upside down and shuffle them.
2. Each student chooses the required number of dominoes to complete the place value mat used for the game (two for 2 whole number places and 2 decimal places; four for 4 whole number places and 4 decimal places).
3. Students arrange their dominoes to form a decimal number with each side of the domino representing a number corresponding to the number of dots on each side.
4. Students write the decimal number they have arranged on the recording sheet.
5. Students compare their numbers with the rest of the group.
6. Student with the highest number is the winner of the round.
7. All dominoes are placed back in the pile and shuffled again for a new round.

DIFFERENTIATION
Add greater decimal values to increase the difficulty level.

Who's in Whose Place?
WHOLE NUMBER MAT

Shirley Disseler   *Strategies and Activities for Common Core Math: Grades 3–5*   Part 1

Who's in Whose Place?
WHOLE NUMBER RECORDING SHEET

Name: _____ Date: _____

Ten Millions	Millions	Hundred Thousands	Ten Thousands	Thousands	Hundreds	Tens	Ones

Who's in Whose Place?
DECIMAL NUMBER MAT

Shirley Disseler — Strategies and Activities for Common Core Math: Grades 3–5 — Part 1 — 101

Who's in Whose Place?
DECIMAL NUMBER RECORDING SHEET

Name: _____ Date: _____

Thou-sands	Hundreds	Tens	Ones	.	Tenths	Hun-dredths	Thou-sandths	Ten Thou-sandths
				.				
				.				
				.				
				.				
				.				
				.				
				.				
				.				
				.				
				.				
				.				

CARD SHARKS

GRADE 3

COMMON CORE STANDARD
3.NBT.A.3

MATHEMATICAL PRACTICES
MP1
MP2
MP6
MP8

MATERIALS
• Dice (one die per pair)
• Card Sharks cards (one set per pair)
• Card Sharks recording sheet (one per student)

OVERVIEW

This game offers **practice multiplying using place value and understanding the role of zero in multiplying by tens**.

PROCEDURE
Game is played in pairs

1. Players lay the three Card Sharks cards face down and mix them around on the desk.
2. Players turn over one card for the round.
3. In turn, each player rolls the die in each round.
4. Each player records his/her problem (the number rolled multiplied by the card overturned in that round) and the solution.
5. At the end of 5 rounds, players add the sum of the solutions. Player with the highest total is the winner.

Shirley Disseler — Strategies and Activities for Common Core Math: Grades 3–5

Card Sharks
CARD SET

1

10

100

Card Sharks
RECORDING SHEET

Name: _____ Date: _____

Example:
Roll: 5
Card turned: 10

Round	Problem	Solution
Example	5 x 10	50
1		
2		
3		
4		
5		
	Total:	

Shirley Disseler — Strategies and Activities for Common Core Math: Grades 3–5

RACE TO THE PLACE

GRADE 4

COMMON CORE STANDARDS
4.NBT.A.2
4.NBT.A.3

MATHEMATICAL PRACTICES
MP6
MP7

MATERIALS

- Scissors (one per student)
- Number cards 0 – 9 (one set per student)
- Notebook paper (one per student)
- Race to the Place recording sheet (one per student)
- Race to the Place assessment sheet (five per student)

OVERVIEW

This game provides students with a different approach to **working with place value**. Paper folding has been proven effective in brain research as a strategy for note taking and practicing math terms and concepts. This activity encourages students to compare the digits in the same place and in different places within the same number.

PROCEDURE

Game is played in pairs

1 Each student folds one piece of notebook paper lengthwise and cuts six slits to the fold to make seven flaps.

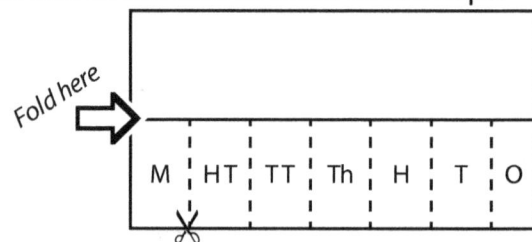

2 Students write the place value on each flap (**O**nes, **T**ens, **H**undreds, **Th**ousands, **T**en **T**housands, **H**undred **T**housands, **M**illions).

3 Students shuffle both sets of number cards together and place the pile face down in the middle of the table.

4 In turn, each player draws one card and places it face down inside the place value flap they want it to represent.

5 When all the places have a card for both students, they turn over the card in each flap without reordering them and record their numbers on their recording sheets.

6 The player with the highest number wins the round.

7 Cards are shuffled and play begins again.

8 After five rounds are played, students complete the assessment sheet.

DIFFERENTIATION

Players can decide before beginning each round if the winning number will be the highest or lowest number.

Race to the Place
RECORDING SHEET

Name: _____ Date: _____

Millions	Hundred Thousands	Ten Thousands	Thousands	Hundreds	Tens	Ones

Race to the Place
ASSESSMENT SHEET

Name: _____ Date: _____

Standard Form _____

Expanded Form _____

Word Form _____

Sum of the digits _____

Total:

Millions _____

Hundred-Thousands _____

Ten-Thousands _____

Thousands _____

Hundreds _____

Tens _____

Ones _____

Rounded to:

Ten _____

Hundred _____

Thousand _____

Hundred-Thousand _____

AREA MODEL MATCH-UP

GRADES 4, 5

COMMON CORE STANDARDS
4.NBT.A.2
4.NBT.B.5
4.NBT.B.6
5.NBT.B.5

MATHEMATICAL PRACTICES
MP1
MP3
MP4
MP5
MP7

MATERIALS

- Base ten block sets (one set per student)
- Area Model Match-Up problem cards (one set per group)
- Timer (one per group)
- Area Model Match-Up recording sheet (one or more per student)
- Modeling with Base Ten Blocks chart

OVERVIEW

This activity helps students **understand two-digit multiplication using arrays and base ten blocks**. This process provides the opportunity for students to make connections between the standard algorithm and rectangular arrays. It also **expands student understanding of place value** in terms of multiplication and breaking apart more difficult problems. This activity helps students see the partial products within a problem as decomposition of the numbers within a problem.

For teachers and students who have not had experience modeling with arrays and base ten blocks, a chart showing the step-by-step procedure of how to model with base ten blocks is included.

PROCEDURE
Whole-class practice

1 Create several practice problems using multiplication of 2 two-digit numbers. Do not duplicate the problems on the Area Model Match-Up problem cards.

2 Have students practice modeling problems using base ten block sets. On the recording sheet, students write the problem, make a sketch of their model, and write the expanded or array form of the problem.

PROCEDURE
Groups of 2 – 3

1 Each student chooses a problem card from the pile but does not show the other players the problem.

2 Set a timer (longer at first, shorter for more challenge).

3 Using base ten blocks, each student builds a model of his/her problem within the allotted time. In turn, each student reveals his/her model to the group.

4 Group members examine each model and write the multiplication problem they think each model represents.

5 Group members share/compare their answers. Students defend and explain any discrepancies.

Area Model Match-Up
PROBLEM CARDS

13 × 12	12 × 21	13 × 22
11 × 21	13 × 11	11 × 12
13 × 13	21 × 13	16 × 17

Area Model Match-Up
PROBLEM CARDS

14 × 16	19 × 21	18 × 19
15 × 13	15 × 17	23 × 14
19 × 12	14 × 21	20 × 15

Area Model Match-Up
RECORDING SHEET

Name: _____ Date: _____

Problem _____	Area model sketch
Expanded form _____	
_____	
_____	
_____	
Answer _____	
_____	

Problem _____	Area model sketch
Expanded form _____	
_____	
_____	
_____	
Answer _____	
_____	

Problem _____	Area model sketch
Expanded form _____	
_____	
_____	
_____	
Answer _____	
_____	

Modeling with Base Ten Blocks

There are four kinds of base ten blocks:

The one block, which is a small cube	The ten rod, which is a long rectangular stick	The flat, which represents 100 cubes or 10 rods	The cube, which is 1000 or ten flats, or 100 rods.
☐		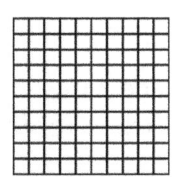	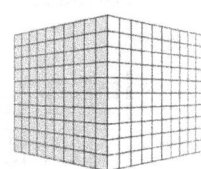

The term "array" refers to the vertical and horizontal multiplication model used to teach basic multiplication, such as 2 sets of 3 or 3 sets of 2.

Using the example *18 x 13*:

The traditional algorithm sets up the problem as

$$\begin{array}{r} 18 \\ \times\ 13 \\ \hline \end{array}$$

Breaking down the problem:

3 x 8 = 24
3 x 10 = 30
10 x 8 = 80
10 x 10 = 100

How to model this problem using base ten blocks:

1 Draw a right angle base. Place the problem, broken down into place value parts, along the outside edges of the angle (*18* down one side, *13* across the other).

2 Begin at the bottom left corner and place one base ten flat in the corner to represent 10 x 10.

Shirley Disseler   Strategies and Activities for Common Core Math: Grades 3–5   Part 1   **113**

3 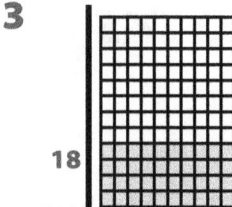 Place 8 base ten rods above the 100 block to represent *10 x 8*.

4 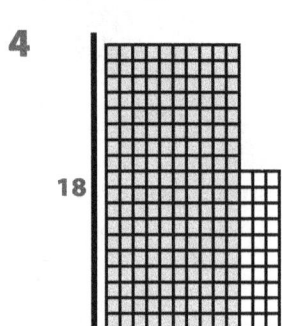 Place 3 base ten rods to the right of the hundreds block to represent *3 x 10*.

5 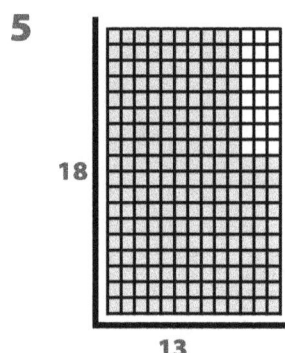 Place 24 single base ten cubes in the angle created above the 30 and to the right of the 80 to represent *3 x 8*. This forms a rectangular array and demonstrates the answer in place value or expanded form.

NOTE
This process works well to model multiplication problems that are two-digit by two-digit numbers.

NUMBER LOGIC IN A BOX

GRADES 4, 5

COMMON CORE STANDARDS
4.OA.C.5
4.NBT.B.4
5.NBT.B.5
5.NBT.B.6

MATHEMATICAL PRACTICES
MP1
MP3
MP7
MP8

MATERIALS
• Number Logic in a Box problems (one per student, pair, or display for whole class)

OVERVIEW

This activity is based on number theory. It asks students to think about **number relationships** in order to find missing numbers in the box. It also involves **computational skills** of addition, subtraction, multiplication, and division.

PROCEDURE

Activity is done individually, in pairs, or with the whole class

1 Students examine each Number Logic box, trying to determine what number belongs in the center blank. Next to each box, students write the reasoning behind their solutions.

2 Have students share their strategies and solutions.

NOTE

More than one solution is permissible—what's important is that the reasoning for the solution makes sense.

Shirley Disseler

Number Logic in a Box
PROBLEMS

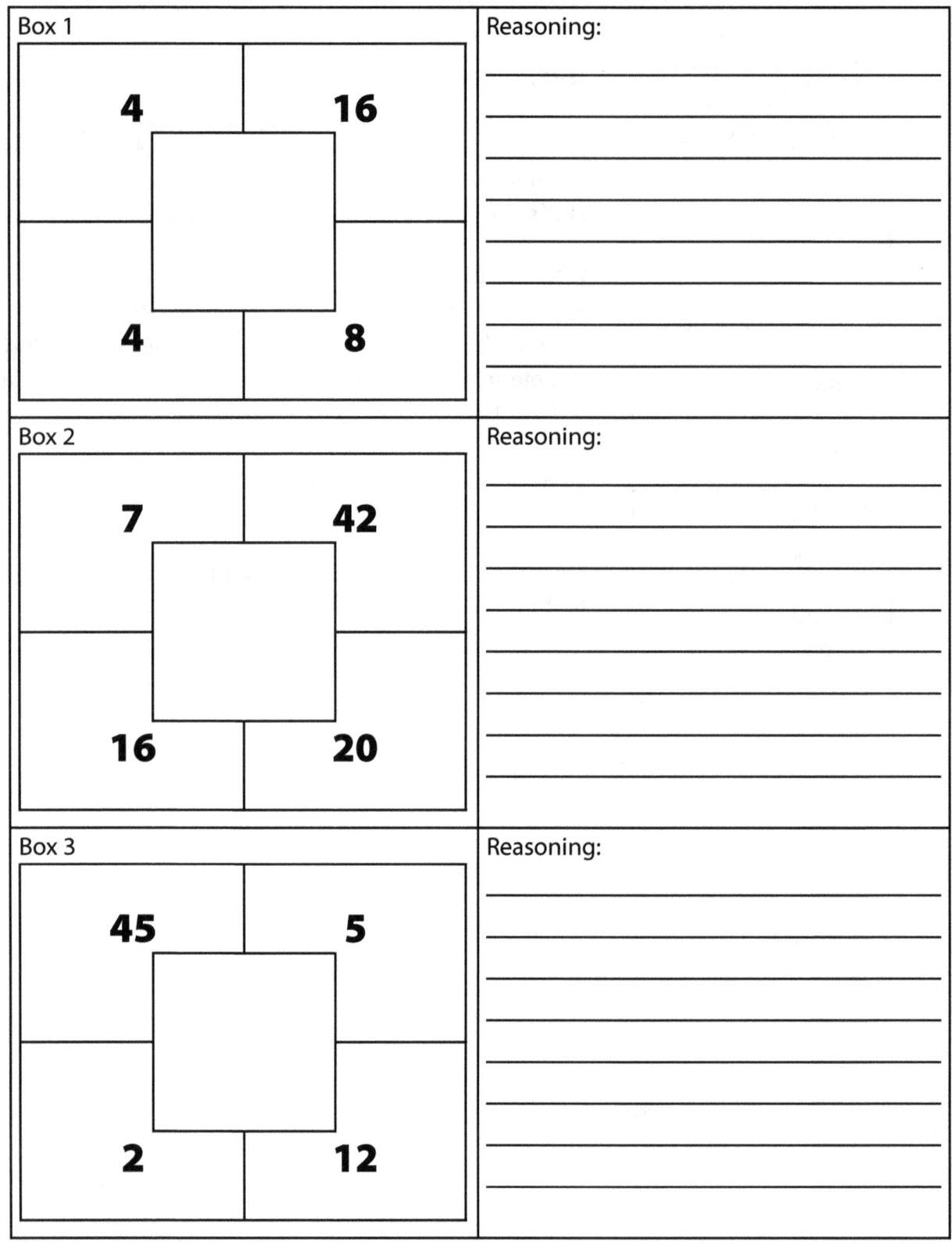

Number Logic in a Box
PROBLEMS

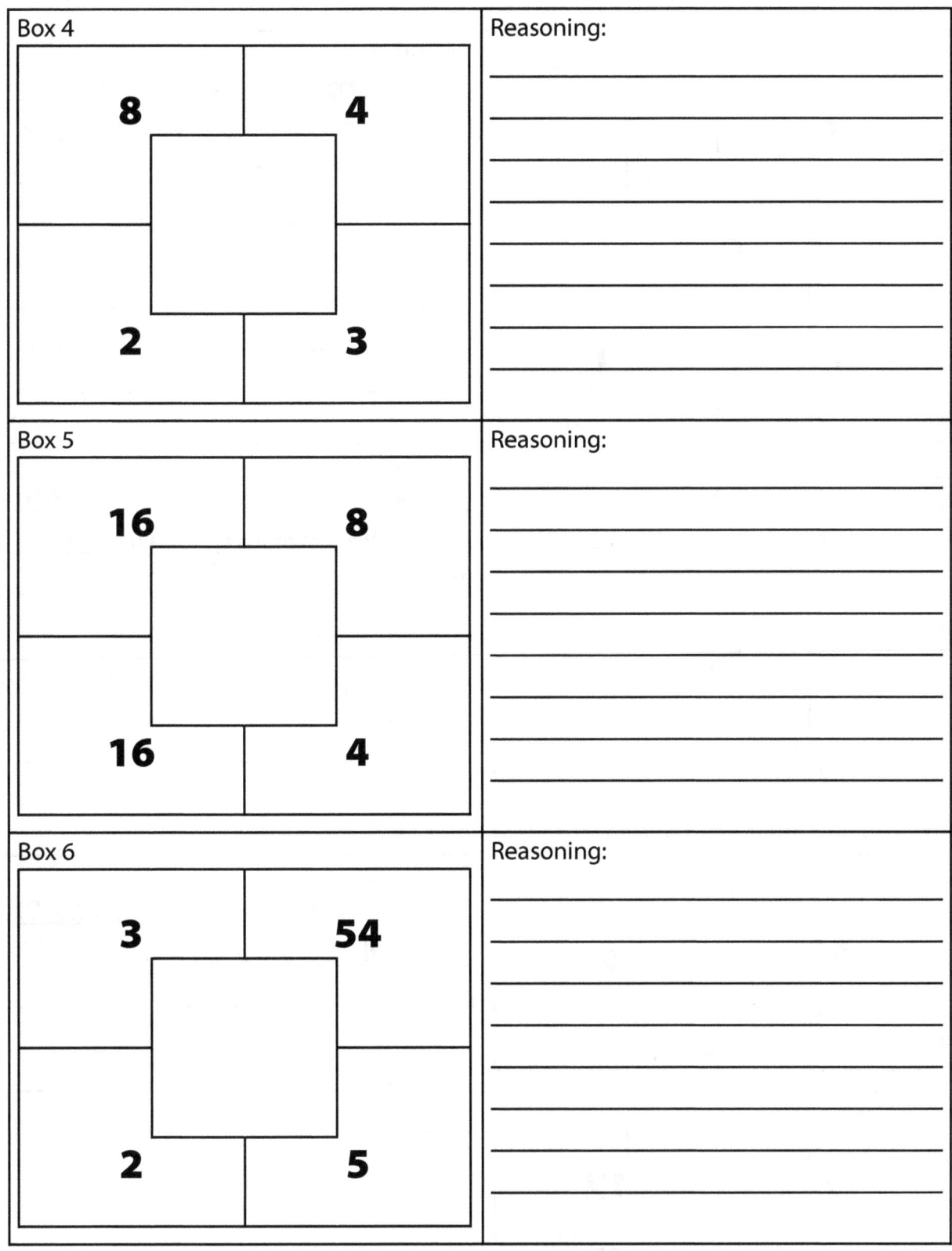

Box 4: 8, 4, 2, 3

Reasoning:

Box 5: 16, 8, 16, 4

Reasoning:

Box 6: 3, 54, 2, 5

Reasoning:

Shirley Disseler — Strategies and Activities for Common Core Math: Grades 3–5

ANSWER SHEET

Number Logic in a Box
POSSIBLE SOLUTIONS

Box 1

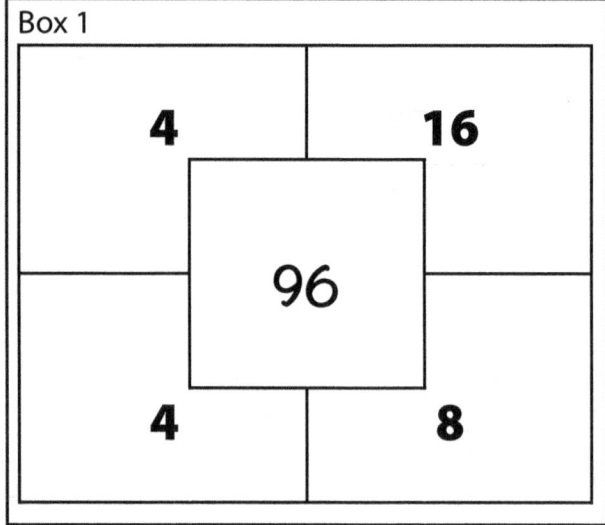

Reasoning:
The top numbers multiplied together (4 x 16 = 64) and added to the product of the bottom numbers (4 x 8 = 32)

Box 2

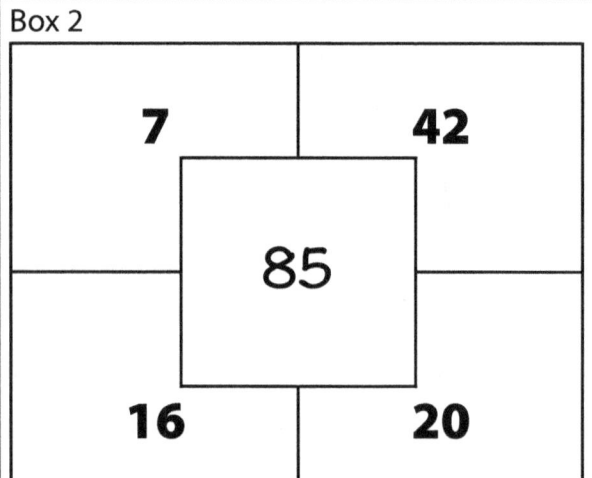

Reasoning:
The sum of the top numbers (7 + 42 = 49) added to the sum of the bottom numbers (16 + 20 = 36)

Box 3

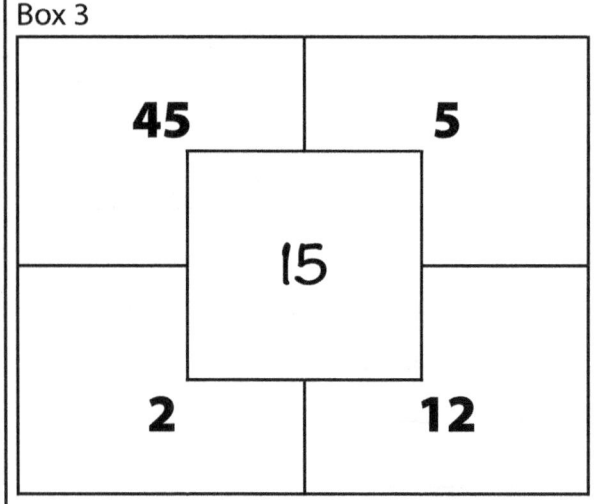

Reasoning:
The quotient of the top numbers (45 ÷ 5 = 9) added to the quotient of the bottom numbers (12 ÷ 2 = 6)

ANSWER SHEET

Number Logic in a Box
POSSIBLE SOLUTIONS

Box 4

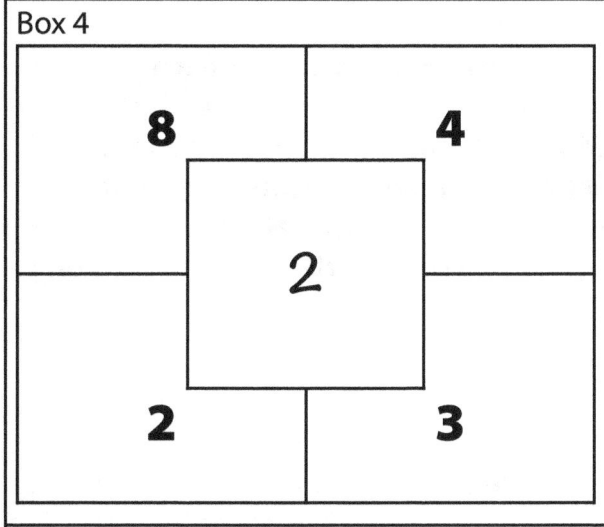

Reasoning:
The difference between the product of the bottom numbers (2 x 3 = 6) and the difference of the top numbers (8 - 4 = 4)

Box 5

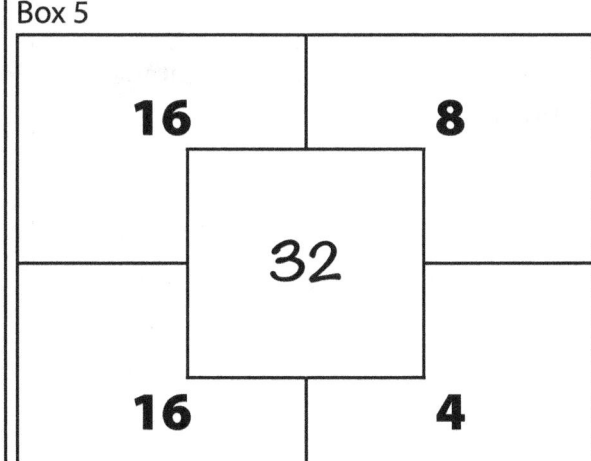

Reasoning:
The sum of the numbers on the left (16 + 16 = 32) is equal to the product of the numbers on the right (8 x 4 = 32)

Box 6

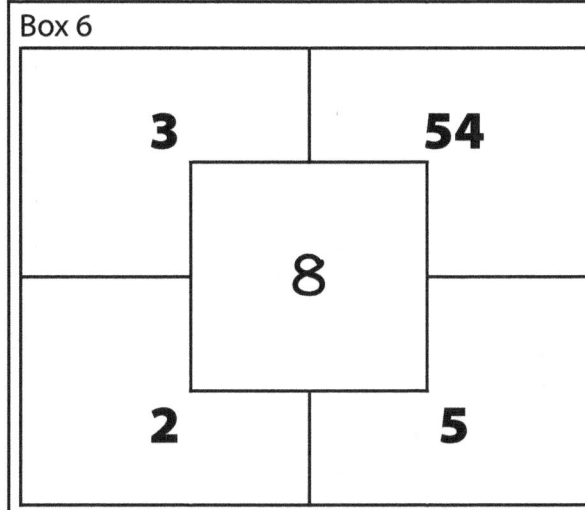

Reasoning:
The difference between the quotient of the top numbers (18) and the product of the bottom numbers (10)

Shirley Disseler — Strategies and Activities for Common Core Math: Grades 3–5 — Part 1   119

PLACE THE NUMBER

GRADE 5

COMMON CORE STANDARDS
5.NBT.A.1
5.NBT.A.2
5.NBT.A.3
5.NBT.B.7

MATHEMATICAL PRACTICES
MP4
MP6
MP7

MATERIALS
• Place the Number clues (one set of clues per student or pair of students)
• Place the Number cards (one set per student or pair of students)
• Place the Number recording sheet (one per student or pair of students)
• Place the Number place value mat (one per student or pair of students)

OVERVIEW

In grade 5, students need **to understand that each place value is a power of ten**. This activity provides practice with this skill using multi-digit numbers and decimals. They also need **to understand that placement denotes numerical meaning and value based upon the base ten system**. This activity also helps students work with understanding the role of multiplication and division within the decimal system.

PROCEDURE

Activity is done in pairs or individually

1 Students read the Place the Number clues. Using the Place the Number cards, they build each number on the Place the Number place value mat.

2 On the recording sheet, students write each number's standard form, word form, and expanded form.

Place the Number
CLUES

1 Place a 5 in the tenths place. Put a number in the ones place that shows the tenths place times 10.

2 Place a 7 in the hundredths place. Place a number in the ones place that shows 100 times the hundredths place.

3 Place a 6 in the hundreds place, a 0 in the tens place and a 5 in the ones place. Place a number in the tenths place that is equivalent to the hundreds place divided by 1000.

4 Build one number that uses 5 fives that have different values and 2 twos that have different values.

5 Make a number that shows 10 times the digit in the ones place and 1/10 the digit in the ones place using a number of your choice.

6 Build a number that has 8 in the tens place, 7 in the tenths place, 6 in the hundreds place. The ones place is 10 times the tenths place. The hundredths place is 1/1000th the number in the tens place.

7 In the number 44.444, what is the value of digit in the thousandths place?

8 Build a number that shows 6 ones and 234 tens.

9 Build a number that shows 9 in the tenths place and 10 times that amount in the ones place. It has 4 digits and no digits are less than 5.

10 Build a number that shows 12 hundredths, and ten times the tenths place using three digits.

11 Build a number that shows consecutive digits less than 6. This number has three digits to the left of the decimal point and has 5 in the tenths place.

12 Build a number that shows eight consecutive digits in descending order. Four digits are on each side of the decimal point. The last number is 0.

Place the Number
CARDS

5	2	3	7	6
5	2	3	7	6
5	2	3	7	6
5	2	3	7	6
5	2	3	7	6
5	2	3	7	6

Place the Number
CARDS

4	8	9	0	1
4	8	9	0	1
4	8	9	0	1
4	8	9	0	1
4	8	9	0	1
4	8	9	0	1

PLACE THE NUMBER
PLACE VALUE MAT

Thousands	Hundreds	Tens	Ones	.	Tenths	Hundredths	Thousandths
				.			
				.			
				.			
				.			

Place the Number
RECORDING SHEET

Name: _____ Date: _____

	Standard form	Word form	Expanded form
1			
2			
3			
4			
5			
6			
7			
8			
9			
10			
11			
12			

ANSWER SHEET

Place the Number
RECORDING SHEET

Name: _____ Date: _____

	Standard form	Word form	Expanded form
1	5.5	five and five tenths	5 + 0.5
2	7.07	seven and seven hundredths	7 + .07
3	605.6	six hundred five and six tenths	600 + 0 + 5 + 0.6
4	(answers can vary) 2555.552	two thousand five hundred fifty-five and five hundred fifty-two thousandths	2000 + 500 + 50 + 5 + 0.5 + 0.05 + 0.002
5	(answers can vary) 55.5	Fifty-five and five tenths	50 + 5 + 0.5
6	687.78	Six hundred eighty-seven and seventy-eight hundredths	600 + 80 + 7 + 0.7 + 0.08
7	444.444	Four hundred forty-four and four hundred forty-four thousandths	400 + 40 + 4 + 0.4 + 0.04 + 0.004
8	2,346	Two thousand three hundred forty-six	2000 + 300 + 40 + 6
9	(answers can vary) 89.97	Eighty-nine and ninety-seven hundredths	80 + 9 + 0.9 + .07
10	1.12	One and twelve hundredths	1 + 0.1 + 0.02
11	234.5	Two hundred thirty-four and five tenths	200 + 30 + 4 + 0.5
12	7654.3210	Seven thousand six hundred fifty-four and three thousand two hundred ten ten-thousandths	7000 + 600 + 50 + 4 + 0.3 + 0.02 + 0.001 + 0.0000

NUMBER & OPERATIONS—FRACTIONS

Fractions are everywhere. Many children experience fractions outside the classroom: adults refer to time in "half an hour" or distance as "three-quarters of a mile." If parents sew, do woodworking, or cook, fractions are part of a family's everyday life. Prior experience with fractions will vary from one student to another, so teachers need to assess before starting instruction on fractions.

Within the Common Core State Standards, the background for fractions is well developed. If teachers in grades K – 2 lay the foundation through sorting, classifying, telling time, money, and partitioning shapes, teachers in grades 3 – 5 have a better chance to help students develop true understanding of fractions.

Terminology is an important part of teaching fractions. Such terms as partitioning, equivalence, decimal fractions, and unit fractions are introduced in the Common Core Standards in grades 3 – 5.

There has been a great of research about the difficulties students have with fractions (Van deWalle, 2013; Cramer & Whitney, 2010; Lamon, 2012). Understanding some of the reasons why can be helpful when teaching:

- Students misunderstand the meaning of the numerator and the denominator. They often see them as separate values and not as a single entity. Using number line activities and rulers will help.

- Students do not interpret fractions as equal parts of a whole. They often define a fraction such as 2/5 as two parts out of 5, rather than equal-sized portions of a whole. Shading and matching shaded parts can help.

- Students lack understanding about size. They think that a fraction such as 1/5 is smaller than 1/10 because 5 is smaller than 10. Visuals are definitely required to solve this problem for students.

- Students use whole number rules for fraction operations, which will not always work. They add or subtract across the top and the bottom, for example. Manipulative materials help make the concept of fractions clearer, as does introducing fractions in a real-world context.

Before grade 6, students should develop skills in all four operations with fractions (addition, subtraction, multiplication, and division) and be able to justify drawings using these operations. Students should be able to read word problems with fractions and utilize strategies to define their solutions.

The activities in this chapter use a number of hands-on strategies to help students understand the concept of fractions. Several activities make the concept of fractions more concrete through the use of measurement tools and money.

PLAY YOUR CARD

GRADES 3, 4

COMMON CORE STANDARDS
3.NF.A.1
3.NF.A.2
4.NF.A.1
4.NF.A.2

MATHEMATICAL PRACTICES
MP1
MP4
MP5
MP7

MATERIALS
- Playing cards (one deck per group)
- Play Your Card mat (one per student)
- Play Your Card scoring sheet (one per student)

OVERVIEW

This game is designed to provide students with **practice with fractional representations of numbers**. Students in grades 3 and 4 need to be able to **identify parts of a fraction and place them on a number line**. This activity helps students attain both of these skills.

PROCEDURE
Play in groups of 3 – 4

1. Each player receives two cards and places them face down on the Play Your Card mat, one in the "numerator" position and the other in the "denominator" position.

    NOTE
    Ace = 1   Jack = 11   Queen = 12   King = 13   Joker = wild

2. Each player turns over the denominator card and describes what that number represents (for example, if the number is a 5, the student might say, "this shows there are five parts of the whole that is being represented").

3. Each player turns over the numerator and describes what that number represents (for example, if the number is a 3, the student might say, "this shows three of the five parts in the whole").

4. Each player writes the fraction on a number line to determine which one is largest, if students have not yet covered the skill of finding common denominators. If they have learned this skill, they can use common denominators to determine the order of the fractions.

5. The players order the fractions in the group from largest to smallest and assign point values (largest: 4 points, second largest: 3 points, third largest: 2 points, and smallest: 1 point)

6. Each player completes the assessment sheet as he/she proceeds through the game, as follows:
    a. Players list the numerator and denominator of their own fraction.
    b. In column 3, players record the fraction they made.
    c. In column 4, players list all the fractions made by the group, in order, and circle their own.
    d. In column 5, players record the points they receive for the round.
    e. Players draw a number line and place each of their fractions from column 3 on the number line.
    f. Players write a word problem using their fractions from column 3.

7. Five rounds are played and points are tallied. The player with the highest score wins.

DIFFERENTIATION
Remove the jacks, queens, and kings from the deck for an easier game.

Play Your Card
GAME MAT

Numerator	Numerator
Denominator	Denominator

Play Your Card
SCORING SHEET

Numerator	Denominator	Fraction	Group's Fractions	Points
			Total Points:	

Number line model (draw a number line and place each fraction on it):

Play Your Card

Name: _____ Date: _____

Write a word problem for each of your fractions to show that you understand what the numerator means and what the denominator means.

Round 1

_____
_____
_____
_____

Round 2

_____
_____
_____
_____

Round 3

_____
_____
_____
_____

Round 4

_____
_____
_____
_____

Round 5

_____
_____
_____
_____

EQUIVALENT FRACTION FRENZY

GRADES 3, 4

COMMON CORE STANDARDS
3.NF.A.1
3.NF.A.2
4.NF.A.1
4.NF.A.2

MATHEMATICAL PRACTICES
MP4
MP6
MP7

MATERIALS
• Equivalent Fraction Frenzy picture cards (one set per group)
• Equivalent Fraction Frenzy recording sheet (one per student)

OVERVIEW

This game helps students practice the skill of **determining equivalency using fractions**. It is designed for grade 4 Common Core Standards, but if students in grade 3 are able to determine equivalency, then it is appropriate to utilize from grade 3. The grade 3 Standards apply if a number line is used to determine equivalency.

PROCEDURE
Game is played in groups of 3

1 Shuffle the Equivalent Fraction Frenzy picture cards and deal each student four cards.

2 Make a pile of the remaining cards and turn the pile upside down in the center of the table. Turn over the top card to start a discard pile.

3 Students play in turns (choose a method to determine who goes first; one good method is to have the youngest student goes first).

4 The first player draws one card and discards one card, trying to make a pair of equivalent fractions.

5 If the student has a pair of equivalent fractions, he/she says, "match," and lays the cards down in front of him/her. Then the matched set is recorded on the student's recording sheet in the two fraction columns.

6 In the last column of the recording sheet, the student explains in writing or mathematically why the two fractions are equivalent.

7 Once all the cards are used, students check who has the most sets. The discard pile can be reshuffled and reused until all sets are completed.

8 Game is over when all sets are completed or until teacher calls time.

Equivalent Fraction Frenzy
PICTURE CARDS

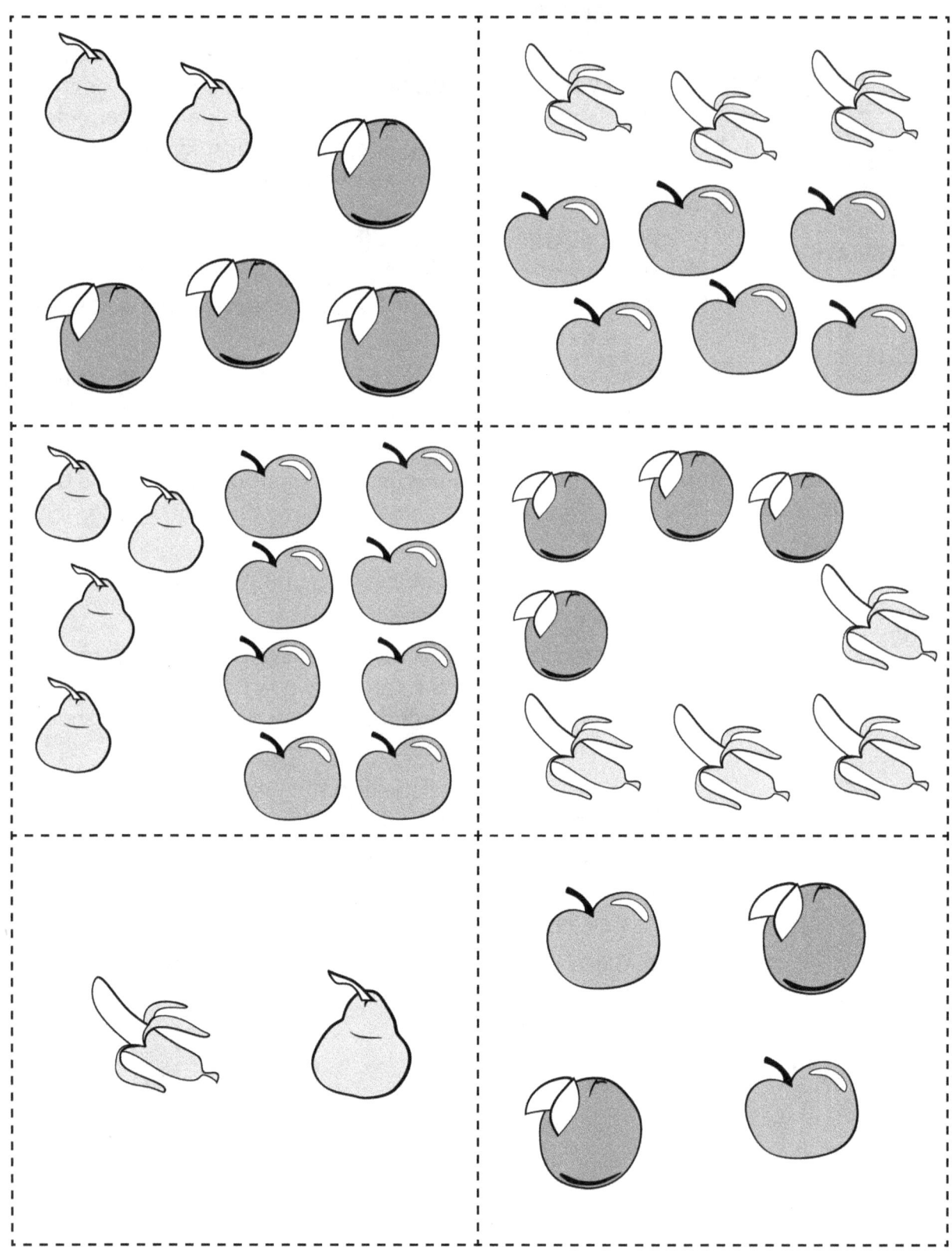

134   Strategies and Activities for Common Core Math: Grades 3–5   Part 1   Shirley Disseler

Equivalent Fraction Frenzy
PICTURE CARDS

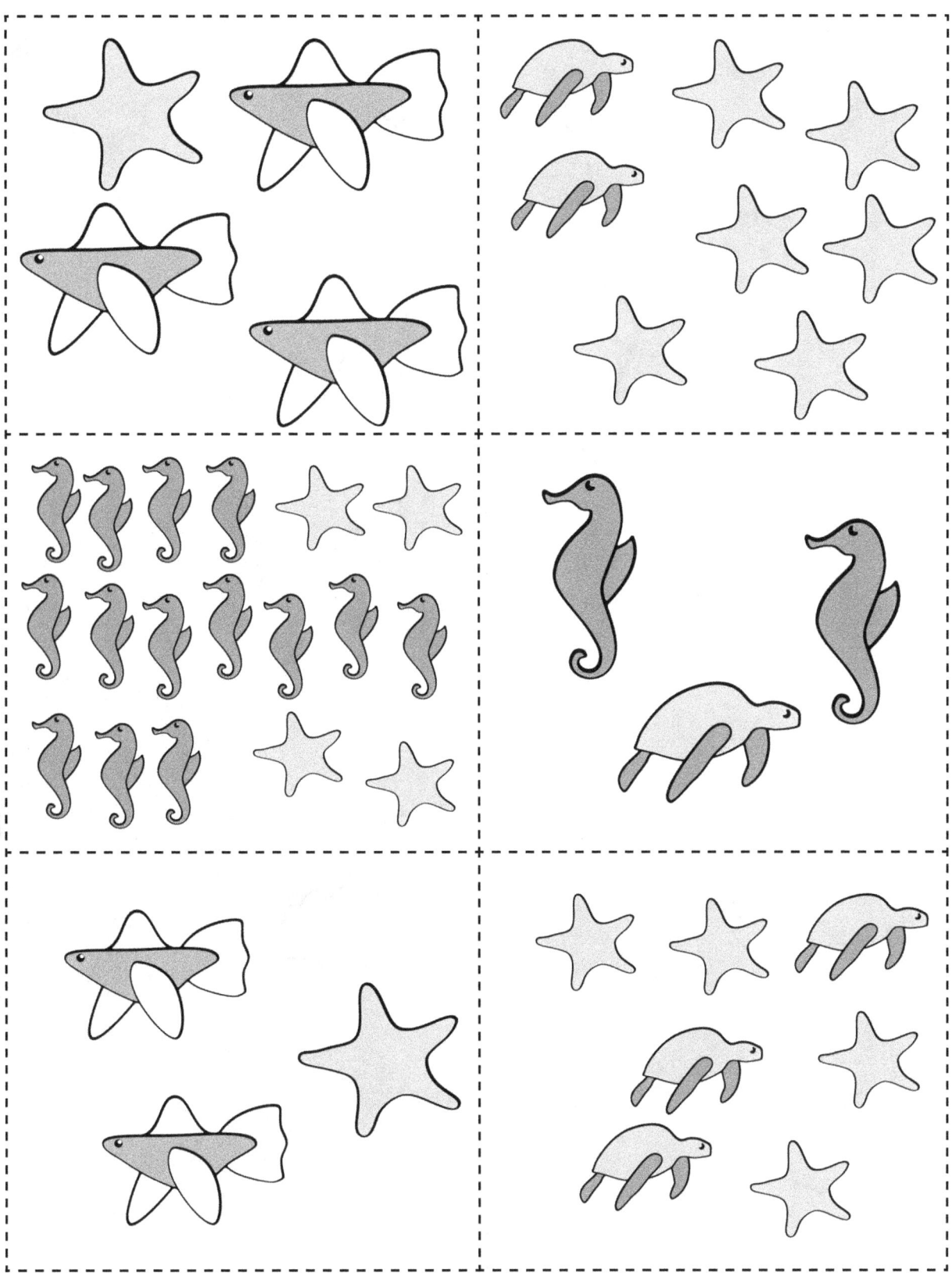

Shirley Disseler — Strategies and Activities for Common Core Math: Grades 3–5 — Part 1   **135**

Equivalent Fraction Frenzy
PICTURE CARDS

136　　Strategies and Activities for Common Core Math: Grades 3–5　　　Part 1　　　Shirley Disseler

Equivalent Fraction Frenzy
RECORDING SHEET

Name: _____ Date: _____

Fraction 1	Fraction 2	Explain why the two fractions are equivalent

Shirley Disseler   Strategies and Activities for Common Core Math: Grades 3–5

FRACTIONS RULE!

GRADES 3, 4, 5

COMMON CORE STANDARDS
3.NF.A.1
3.NF.A.3
4.NF.A.1
4.NF.A.2
5.NF.A.1
5.NF.A.2

MATHEMATICAL PRACTICES
MP3
MP4
MP6

MATERIALS
• Colored tokens (use Skittles® if your school doesn't object)
• Fractions Rule! page (one per student)

OVERVIEW

This group activity **combines student understanding of measurement**—reading a ruler—**with fractional representations in a model format**. It is important that students see fractions in a real-world context, such as a ruler, and understand that they are utilized in many areas of life.

PROCEDURE

Activity is conducted in groups of 3

1 Using the tokens, students cover the marks on the Fractions Rule! activity pages based on the rule given.

2 Students compare with other students in their group and defend their placements of the tokens.

3 Students write statements about why they covered the marks as they did.

Fractions Rule!

1. Cover all the marks with a red token that represents fourths.
   Explain why these marks are covered.

   _____
   _____

2. Use a blue token to cover the mark that represents the solution to 4 ⅞ - 2 ¼.
   Explain why this mark is covered.

   _____
   _____

3. Use a green token to cover the fraction that represents the solution to 7 ¼ - 4 ½.
   Explain why this mark is covered.

   _____
   _____

4. Use an orange token to cover all lines that represent ⅛ on the ruler.
   Explain why these marks are covered.

   _____
   _____

5. Cover all the marks that are on the ½ inch lines with any color token. Explain why these marks are covered.

_____

_____

6. Use a yellow token to cover all multiples of ⅜. Explain why these marks are covered.

_____

_____

MEET ME IN THE MIDDLE

GRADES 3, 4, 5

COMMON CORE STANDARDS
3.NF.A.3
4.NF.A.2
5.NF.A.1

MATHEMATICAL PRACTICES
MP1
MP3
MP6

MATERIALS
• Meet Me in the Middle 1 – 9 cards (one set per student)—can also use ace – 9 playing cards
• Meet Me in the Middle mat (one per student)
• Small tokens (ten per student – use Skittles® if your school doesn't object)
• Meet Me in the Middle recording sheet (one per student)

OVERVIEW

This game is a **review strategy for comparing fractions**. Students will determine the order of fractions by **finding common denominators or utilizing other fraction rules**.

PROCEDURE

Game is played in pairs.

1. Each player shuffles his/her set of cards.
2. Each player places the first three cards in the numerator spots on the mat, face down.
3. Each player places the next three cards in the denominator spots on the mat, face down.
4. Players should not look at the three leftover cards.
5. Players turn over all the cards on the mat. Using their tokens, they cover the signs, leaving open the signs that make the statements true.
6. On the recording sheet, players record the numbers, signs, and explanations of the statements.

SCORING

If the statements can be validated, the player scores 5 points.
If the statements cannot be validated, the player can opt to switch the cards or the tokens on the mat once. Then, if the player can make the statement true, he/she scores 4 points.
If the player cannot make a true statement, he/she receives no points.

DIFFERENTIATION

For a simpler game, compare two fractions.

Shirley Disseler — Strategies and Activities for Common Core Math: Grades 3–5 — Part 1   **141**

Meet Me in the Middle
1 – 9 CARDS

1	2	3
4	5	6
7	8	9

Meet Me in the Middle
MAT

Place the 1 – 9 cards in the boxes to make the statement true. Use tokens to cover the extra signs and leave the signs that make the statement true.

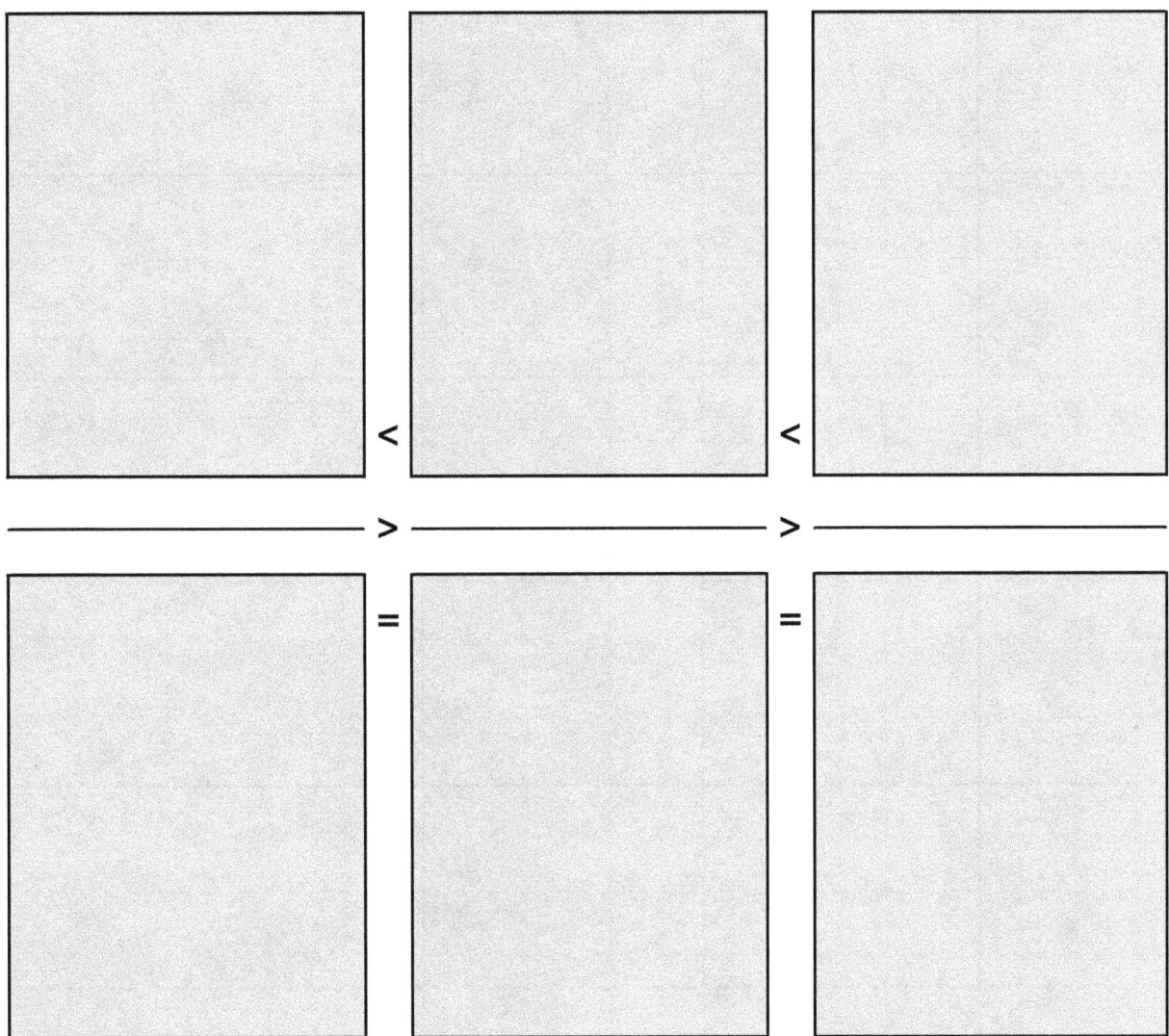

Shirley Disseler — Strategies and Activities for Common Core Math: Grades 3–5 — Part 1 **143**

Meet Me in the Middle
RECORDING SHEET

Fraction 1	Sign	Fraction 2	Sign	Fraction 3	Explain your statement

WHERE DO I BELONG?

GRADE 4

COMMON CORE STANDARDS
4.NF.A.1
4.NF.A.2

MATHEMATICAL PRACTICES
MP2
MP3
MP6
MP8

MATERIALS
• String
• Where Do I Belong? large fraction cards
• Where Do I Belong? fraction circle chart (one per student)
• Where Do I Belong? small fraction cards

OVERVIEW
Whole-class and individual activity

The purpose of this activity is to determine what students know about **equivalent fractions and names for fractional parts**. This activity can be used as a whole-class engagement activity at the beginning of a learning cycle and as a whole-class formative assessment at the end of a learning cycle. Follow up the whole-class activity with one that the students can do individually or in groups.

WHOLE-CLASS PROCEDURE

1. Make two large circles on the floor with string. Circles must be large enough for several students to sit inside at one time.
2. Label each circle with a fraction in simplest form (¼, ⅓, ½, ¾).
3. Give each student a large fraction card.
4. Students determine if they belong in either of the two large circles. If they do, those students sit in that circle.
5. Have a class discussion about the equivalent fractions, with students defending their choices.

INDIVIDUAL FOLLOW-UP ACTIVITY

1. Each student receives a Where Do I Belong? fraction circle chart, with the two circles labeled by the teacher (example: ½ and ⅓).
2. Students cut out fractions from page of small fraction cards and place them into the circle that best meets the label. They place the ones that don't fit in either circle around the outside of the chart.
3. Students explain their placements in a class discussion.

DIFFERENTIATION

For more accelerated students, use decimal equivalents.

Where Do I Belong?
FRACTION CIRCLE CHART

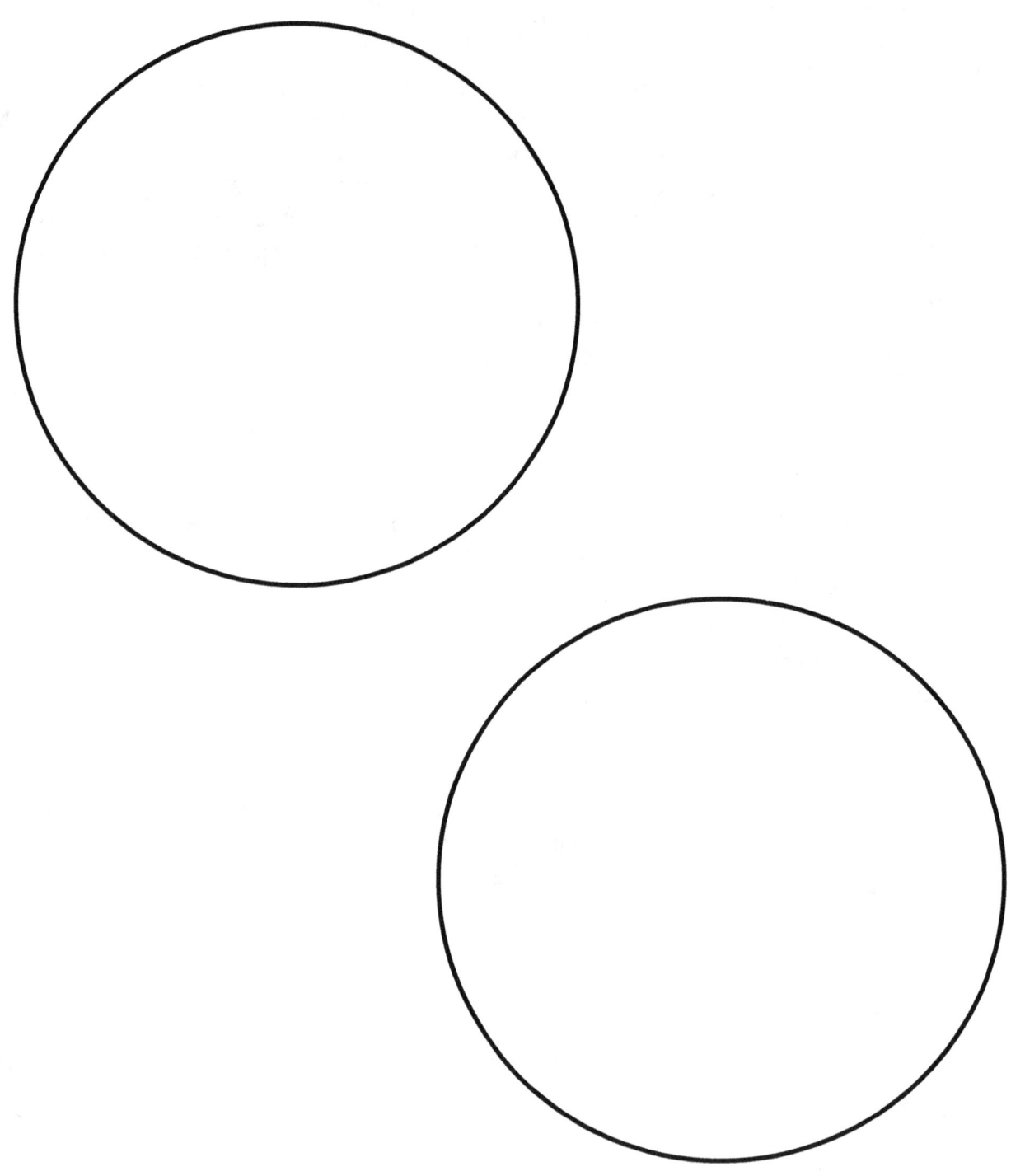

146 *Strategies and Activities for Common Core Math: Grades 3–5* *Part 1* Shirley Disseler

Where Do I Belong?
LARGE FRACTION CARDS

$\frac{4}{8}$	$\frac{5}{10}$	$\frac{4}{16}$
$\frac{6}{12}$	$\frac{3}{18}$	$\frac{3}{9}$
$\frac{3}{15}$	$\frac{7}{14}$	$\frac{2}{10}$

Where Do I Belong?
LARGE FRACTION CARDS

2/8	6/8	2/10
3/12	4/12	5/15
8/12	8/16	8/24

Where Do I Belong?
LARGE FRACTION CARDS

8/24	6/9	5/20
7/21	8/32	6/18
6/24	7/28	8/40

Where Do I Belong?
SMALL FRACTION CARDS

4/8	5/10	4/16	2/8	6/8
6/12	3/18	3/9	2/10	5/15
3/15	7/14	2/10	3/12	4/12
8/12	8/16	8/24	6/9	5/20
7/21	8/32	6/18	6/24	7/28

MAKE A BUCK

GRADES 4, 5

COMMON CORE STANDARDS 4.NF.B.3 5.NF.A.1

MATHEMATICAL PRACTICES MP1 MP3 MP6

MATERIALS • Make a Buck cards (one set per group of 3) • Make a Buck recording sheet (one per student)

OVERVIEW

This game **integrates money into the study of fractions**. It also allows for an expansion into the decimal equivalents of fractions. Addition of fractions and decimals is utilized. This game is also applicable to decimals if students record each fraction as a decimal amount. Practicing decimals as money is a good way to use real-world connections.

PROCEDURE

Game is played in groups of 3

1 Set a time limit for the game (ten minutes works well for most groups).

2 Shuffle the Make a Buck cards and deal four cards to each player.

3 In turn, players draw one card and discard one card onto a discard pile, trying to accumulate cards that add up to 1.00.

4 When a player has $1.00 equivalent, he/she calls out, "I've got a buck!"

Example
With the cards ½ and ¾ in his/her hand, a player would have the equivalent of $1.00, and would call out, "I've got a buck!"

5 When a player calls out, "I've got a buck!" all players check to see if they agree.

6 If they do, the round is over. The cards are returned to the deck and reshuffled for round 2.

7 Players complete the recording sheets for fractional and decimal representations each round.

8 The game ends when the time is called.

Make a Buck
CARDS

½	⅔	⅘
⅓	2/4	4/6
¼	⅕	¾

Make a Buck
CARDS

3/5	2/6	3/6
2/8	4/8	6/8
3/9	1/2	1/4

Make a Buck
CARDS

3/4	1/10	1/10
1/4	5/10	3/6
1/10	2/10	3/10

Make a Buck
CARDS

| 5/10 | 8/10 | 9/10 |

Shirley Disseler — *Strategies and Activities for Common Core Math: Grades 3–5*

Make a Buck
RECORDING SHEET

Name: _____ Date: _____

Fractional Representation

Card drawn	Addition problem	Solution

Decimal Representation

Cards drawn	Decimal values	Addition problem	Dollar amount solution

ROLLING INTO FRACTIONS

GRADES 4, 5

COMMON CORE STANDARDS
4.NF.B.3
5.NF.A.1
5.NF.B.3
5.NF.B.4

MATHEMATICAL PRACTICES
MP1
MP6
MP7
MP8

MATERIALS
- Four dice (two white and two of another color) for each pair of students
- Rolling into Fractions recording sheet (one per student)

OVERVIEW

This game provides practice with the skills necessary for **computing with fractions**. Students use discovery to determine strategies when using addition, subtraction, multiplication, and/or division.

PROCEDURE
Game is played in pairs

1. Give each pair a set of four dice (two white and two of another color). The white dice represent the numerator and the other color dice represent the denominator.

2. Player 1 rolls all four dice and writes the numbers in the boxes on the assessment sheet.

3. Player 1 can choose to add, subtract, multiply, or divide the two fractions. Player 1 writes the chosen symbol in the math symbol box.

4. Player 2 repeats the same process.

5. Both players compute the fractional algorithm and compare answers.

6. The player whose answer is the higher number wins the round.

Rolling into Fractions
RECORDING SHEET

Name: _____ Date: _____

Round	Fraction 1	Fraction 2	Math symbol	Computation work *(show all work here)*	Solution
1					
2					
3					
4					
5					

HOW DOES YOUR GARDEN GROW?

GRADES 4, 5

COMMON CORE STANDARDS
4.NF.B.3
4.NF.C.5
4.NF.C.6
4.NF.C.7
5.NF.A.1
5.NF.A.2
5.NF.B.4
5.NBT.A.1

MATHEMATICAL PRACTICES
MP1
MP2
MP3
MP6
MP7

MATERIALS

- Colored pencils (one set per team)
- How Does Your Garden Grow? garden diagram (one per team)
- How Does Your Garden Grow? questions (one set per team)

OVERVIEW

This activity is designed to challenge students to **use measurement and fractional parts together in a real world problem-solving scenario**. It includes decimals and percentages, which require higher-level thinking skills. This activity is to be used as a **challenge activity**.

PROCEDURE

Activity is done in groups of 3 – 4

1 Using the garden diagram, students color the garden according to the key.

2 Using the garden diagram, students work together to answer the How Does Your Garden Grow? questions.

3 After all the groups have finished, students compare answers and discuss differences.

How Does Your Garden Grow?
GARDEN DIAGRAM

Name: _____ Date: _____

Each block is 15 meters long and 15 meters wide

Corn	Corn	Beans	Pumpkins	Wheat
Corn	Corn	Beans	Pumpkins	Wheat
Corn	Corn	Tomatoes	Peas	Wheat
Corn	Corn	Tomatoes	Peas	Wheat
Corn	Corn	Tomatoes	Onions	Wheat

Corn = yellow
Beans = green
Tomatoes = red
Cucumbers = blue

Pumpkins = orange
Peas = gray (lightly color black)
Onions = pink
Wheat = brown

How Does Your Garden Grow?
QUESTIONS: PART 1

Name: _____ Date: _____

Calculate the dimensions of the garden:

1. Length of the garden: _____
2. Width of the garden: _____
3. One section of the garden: _____
4. Entire garden: _____

Find the area of each crop in square meters.

Crop	Area in square meters
Corn	
Beans	
Tomatoes	
Cucumbers	
Pumpkins	
Peas	
Onions	
Wheat	

How Does Your Garden Grow?
QUESTIONS: PART 2

Name: _____ Date: _____

Find the value of each crop in decimals, fractions, and percentages:

Crop	Fractional value	Decimal value	Percentage
Corn			
Beans			
Tomatoes			
Cucumbers			
Pumpkins			
Peas			
Onions			
Wheat			

HINTS
- Measurement of area: A = L x W
- Denominator: total number of sections in the garden
- Simplify fractions to lowest terms
- Decimals: Divide numerator by denominator
- Percentage: Move the decimal two places

How Does Your Garden Grow?
QUESTIONS: PART 1

ANSWER SHEET

Name: _____ Date: _____

Calculate the measurement and area of the garden as follows:

1. Length of the garden: **75 meters**
2. Width of the garden: **75 meters**
3. One section of the garden: **225 meters squared**
4. Entire garden: **5,625 meters squared**

Find the area of each crop in square meters.

Crop	Area in square meters
Corn	10 plots of corn at 225 square meters each equals 2250 meters squared
Beans	2 plots of beans at 225 square meters each equals 450 meters squared
Tomatoes	2 plots of tomatoes at 225 square meters each equals 450 meters squared
Cucumbers	1 plot of cucumbers equals 225 meters squared
Pumpkins	2 plots of pumpkins at 225 square meters each equals 450 meters squared
Peas	2 plots of peas at 225 square meters each equals 450 meters squared
Onions	1 plot of onions equals 225 meters squared
Wheat	5 plots of wheat at 225 square meters each equals 1,125 meters squared

How Does Your Garden Grow?
QUESTIONS: PART 2

Name: _____ Date: _____

Find the value of each crop in decimals, fractions, and percentages:

Plant	Fractional value	Decimal value	Percentage
Corn	2/5 of the garden	0.4	40%
Beans	2/25	0.08	8%
Tomatoes	2/25	0.08	8%
Cucumbers	1/25	0.04	4%
Pumpkins	2/25	0.08	8%
Peas	2/25	0.08	8%
Onions	1/25	0.04	4%
Wheat	1/5	0.2	20%

HINTS
Measurement of area: A = L x W
Denominator: total number of sections in the garden
Simplify fractions to lowest terms
Decimals: Divide numerator by denominator
Percentage: Move the decimal two places

THE GREAT FRACTION HUNT

GRADES 4, 5

COMMON CORE STANDARDS
4.NF.B.3
4.NF.B.4
5.NF.A.1
5.NF.B.3
5.NF.B.7

MATHEMATICAL PRACTICES
MP1
MP2

MATERIALS

- The Great Fraction Hunt clue cards (one set)
- The Great Fraction Hunt fraction cards (one set)

OVERVIEW

The purpose of this whole-class activity is to **assess the degree to which students understand fractions in terms of greater than and less than, addition and subtraction, and problem solving.**

PROCEDURE

1. Divide the class in two equal groups. Give half the students clue cards and the other half fraction cards.

   NOTE
   There are fraction and clue cards for a class of 26. Create more if needed, so each student has one clue or fraction card.

2. Students with fraction cards line up on one side of the room. Students with clue cards line up on the other side of the room.

3. Without discussing, students walk around the room to find their match.

   NOTE
   You can play music for a specific amount of time and have the students move until the music stops.

4. Check to see who has a correct match and continue until everyone is matched and can explain why.

The Great Fraction Hunt
CLUE CARDS

The fraction that is greater than ¼ but smaller than ½	The fraction that shows the equivalent to ½ of 12	A fraction that is the same as ⁶⁄₈
The fraction that is equivalent to ¼ of 1½	A fraction that is greater than ½ but less than ¾	The fraction that is equivalent to ⅔ of 12
A fraction that is the same as ²⁴⁄₄₈	The fraction that lies halfway between 3½ and 4	The simplest form for ¹²⁄₈
The simplest form for ¹⁶⁄₃	The fraction that shows ⅔ ÷ ¼	The fraction that shows ½ ÷ ⅘
	The fraction that shows ⅔ x ⅗	

The Great Fraction Hunt
FRACTION CARDS

3/8	6/1	3/4
3/8	2/3	8/1
1/2	3 3/4	1 1/2
5 1/3	5/8	6/15
	2 2/3	

Shirley Disseler — *Strategies and Activities for Common Core Math: Grades 3–5*

DECIMALS AND FRACTIONS IN ACTION

GRADE 4

COMMON CORE STANDARD
4.NF.C.6

MATHEMATICAL PRACTICES
MP1
MP6

MATERIALS

- Decimals and Fractions in Action decimal cards (one set per group)
- Decimals and Fractions in Action fraction strips (one set per group)
- Decimals and Fractions in Action recording sheet (one per student)

OVERVIEW

The purpose of this game is to **assess student understanding of both fractional and decimal representations of fractions**. The game helps students **compare numbers written as fractions and those written as decimals.** This game is designed as a challenge for those already have a base knowledge of fractions and decimals.

PROCEDURE

Game is played in groups of 3 – 4

1 Give each group one set of fraction strips (to tenths) and one set of decimal cards.

2 Players shuffle the strips and turn them upside down on the table, spread out between players.

3 Players shuffle and deal four decimal cards to each player.

4 Player 1 draws a fraction strip. If he/she has a matching decimal amount, he/she keeps the strip and places the match on the table.

Example
A match is the fraction strip for thirds and a decimal card of .33 or .66.

5 If the player does not have the match, the strip is returned face down and remixed in the strip pile.

6 Play continues in this fashion until all strips are accounted for.

7 Matches are totaled and the person with the most matches wins the game.

8 Players record matches and explanations on the recording sheet.

DIFFERENTIATION

For more advanced students, use fractions strips to twelfths and create additional decimal cards that correspond to sevenths, eighths, ninths, elevenths, and twelfths.

Decimals and Fractions in Action
DECIMAL CARDS

.10	.80	.20
.70	.30	.60
.40	.50	.90

Decimals and Fractions in Action
DECIMAL CARDS

1.00	1.00	.50
.25	.50	.75
1.00	.33	.66

Decimals and Fractions in Action
DECIMAL CARDS

1.00	.20	.40
.60	.80	1.00
.125	.25	.375

Decimals and Fractions in Action
DECIMAL CARDS

.50	.625	.75
.875	.25	1.00
.166	.667	.833

Decimals and Fractions in Action
FRACTION STRIPS (TO TENTHS)

1

½	½
⅓	⅓	⅓
¼	¼	¼	¼
⅕	⅕	⅕	⅕	⅕
⅙	⅙	⅙	⅙	⅙	⅙
1/10	1/10	1/10	1/10	1/10	1/10	1/10	1/10	1/10	1/10

Decimals and Fractions in Action
FRACTION STRIPS (TO TWELFTHS)

1

| ½ | ½ |

| ⅓ | ⅓ | ⅓ |

| ¼ | ¼ | ¼ | ¼ |

| ⅕ | ⅕ | ⅕ | ⅕ | ⅕ |

| ⅙ | ⅙ | ⅙ | ⅙ | ⅙ | ⅙ |

| ⅐ | ⅐ | ⅐ | ⅐ | ⅐ | ⅐ | ⅐ |

| ⅛ | ⅛ | ⅛ | ⅛ | ⅛ | ⅛ | ⅛ | ⅛ |

| ⅑ | ⅑ | ⅑ | ⅑ | ⅑ | ⅑ | ⅑ | ⅑ | ⅑ |

| ⅒ | ⅒ | ⅒ | ⅒ | ⅒ | ⅒ | ⅒ | ⅒ | ⅒ | ⅒ |

| 1/11 | 1/11 | 1/11 | 1/11 | 1/11 | 1/11 | 1/11 | 1/11 | 1/11 | 1/11 | 1/11 |

| 1/12 | 1/12 | 1/12 | 1/12 | 1/12 | 1/12 | 1/12 | 1/12 | 1/12 | 1/12 | 1/12 | 1/12 |

Decimals and Fractions in Action
RECORDING SHEET

Fraction	Decimal	Explanation

NAME THAT DECIMAL

GRADE 4

COMMON CORE STANDARDS 4.NF.C.6 4.NF.C.7

MATHEMATICAL PRACTICES MP2 MP6 MP7

MATERIALS • Two large dice • Name That Decimal recording sheet

OVERVIEW

This whole-class activity provides **practice using decimal notation to represent fractions**. It involves comparisons of numbers in decimals to the hundredths place.

PROCEDURE

1. For each number, the class decides whether to make the highest or lowest number possible.

2. Teacher rolls the dice and calls out the two numbers rolled.

3. Using pens or markers (not pencils), students write the numbers rolled in the two boxes in #1 on the Name That Decimal recording sheet.

4. Teacher asks: "Who has the highest (or lowest) decimal number?"

5. Students respond with their numbers. Teacher records the numbers on the board. Student must read the number correctly to get credit.

6. If the student has met the criteria, he/she places a check next to the number on the page in the "correct" box. If others have the same number, they also check their number.

7. Teacher repeats the roll again. Student record the numbers in two boxes in #2. Teacher rolls one die, and students fill in the third box in #2.

8. Play continues to end of recording sheet.

DIFFERENTIATION

Students also write each number in word form.

Name That Decimal
RECORDING SHEET

Name: _____ Date: _____

H or L		correct	Word Form
1		☐	_____
2		☐	_____
3		☐	_____
4		☐	_____
5		☐	_____
6		☐	_____
7		☐	_____
8		☐	_____
9		☐	_____

FEATURING FRACTIONS

GRADE 5

COMMON CORE STANDARDS
5.NF.B.5
5.NF.B.6

MATHEMATICAL PRACTICES
MP1
MP2
MP3
MP4
MP7

MATERIALS

- Featuring Fractions numeric problems (one set per group)
- Featuring Fractions word problems (one set per group)
- Featuring Fractions recording sheet (three per group)
- Fraction bars (one set per group)

OVERVIEW

In this activity, students develop models of multiplication using fractional parts. This activity helps students **understand how multiplication affects fractional parts and the comparisons created by multiplying fractions and whole numbers**.

PROCEDURE

Activity is done in groups of 4

1. Each group has one set of numeric problems and one set of word problems. Cut apart the problems. Keep each set of problems separate from each other, but mix each set thoroughly.

2. Students model the problems using fraction bars, to ensure they understand what each word problem means.

3. Students match each numeric problem to a word problem that demonstrates the operation, gluing the matches on the Featuring Fractions recording sheet.

DIFFERENTIATION

Students can draw their models for each problem on the back of the recording sheet.

NOTE

A demonstration of how to use fraction bars to model similar problems can be seen at:
www.youtube.com/watch?v=u0HpqylFv3o

Featuring Fractions
NUMERIC PROBLEMS

⅓ × 5/9	⅓ × 2/7
½ × ⅓	⅛ × ½
4/5 × ⅔	3/5 × ⅓
½ × 2/5	¾ × 2/5
⅔ × 3/5	4/5 × ¼
¼ × ⅔	1/5 × ⅔
6 × ½	5 × 2/5
9 × ⅓	3 × ⅓
12 × 5/6	3 × ⅔
8 × ¼	12 × ¼
4 × ¼	12 × 1/6

Featuring Fractions
WORD PROBLEMS

Tom bought ⅓ of his plants at the garden store. He planted 5/9 of the plants he bought.	⅓ of the class were girls. 2/7 of those wore yellow ribbons in their hair.
Shawn invited ⅓ of his class to the party and ½ of them showed up.	⅛ of the teachers at the middle school were male and ½ of those were married.
⅘ of the dogs in the dog show were labs and ⅔ of the labs were puppies.	⅗ of the baseball team went out for pizza. ⅓ of those had pepperoni pizza.
½ of the boys in the class saw the new movie. ⅖ of those liked it.	¾ of the candy in the bag was chocolate. ⅖ of that candy had peanuts in it.
⅔ of the girls in the class like to swim. ⅗ of those also like to dive.	⅘ of the people on the cruise had their pictures taken. ¼ of those wore evening gowns in their pictures.
¼ of the people at the concert were wearing cowboy boots. ⅔ of those were male.	⅕ of the girls on the tennis team wanted to wear shorts. ⅔ of those wanted green shorts.
There were 6 candy bars on the counter. ½ of them were chocolate peanut bars.	5 people went to dinner. ⅖ of those had steak.
There were 9 people at the scout meeting, but ⅓ left early.	3 children were playing in the park. ⅓ of them wore jackets.
There were 12 puppies to choose from. ⅚ of them were mixed-breeds.	3 men were walking together down the street. ⅔ of them were talking on cell phones.
John had 8 hats, and ¼ of them were baseball hats.	¼ of the people were singing. There were 12 people in the choir.
Shayla had 4 purses and ¼ of them were blue.	12 people went to the show, and ⅙ had popcorn.

Featuring Fractions
RECORDING SHEET

Numeric Problem	Word Problem

Bibliography

Barnett-Clarke, Carne, William Fisher, Rick Marks, and Sharon Ross. *Developing Essential Understanding of Rational Numbers for Teaching Mathematics in Grades 3–5*. Reston, VA: National Council of Teachers of Mathematics, 2010.

Battista, Michael. "The Development of Geometric and Spatial Thinking." In *Second Handbook of Research on Mathematics Teaching and Learning*, edited by F. K. Lester, Jr., 843. Charlotte, NC: Information Age, 2007.

Beckmann, Sybilla (lead writer). *The Mathematical Education of Teachers II: Issues in Mathematics Education Volume 17*. Providence, RI: American Mathematical Society, 2012.

Brophy, Jere. "Fostering Student Learning and Motivation in the Elementary School Classroom." In *Learning and Motivation in the Classroom*, edited by S. Paris, G. Olson, and H. Stevenson, 283-305. Hillsdale, NJ: Lawrence Erlbaum Associates, 1983.

Burns, Marilyn. *About Teaching Mathematics: A K-8 Resource*. Sausalito, CA: Math Solutions Publications, 2007.

Cramer, K. and S. Whitney. "Learning Rational Number Concepts and Skills in Elementary School Classrooms." In *Teaching and Learning Mathematics: Translating Research for Elementary School Teachers,* edited by D. V. Lambin and F. K. Lester, Jr., 15-22. Reston, VA: National Council of Teachers of Mathematics, 2010.

Dweck, Carol S. "Motivational Processes Affecting Learning." *American Psychologist 41*, no. 1 (1986): 31-33.

Garrison, Catherine and Michael Ehringhaus. "Formative and Summative Assessments in the Classroom." www.amle.org/BrowsebyTopic/WhatsNew/WNDet.aspx?ArtMID=888&ArticleID=286

Glaser, Edward. *An Experiment in the Development of Critical Thinking*. New York: Teacher's College, Columbia University, 1941.

Kaftan, J., Gayle Buck, and A. Haack. "Using Formative Assessment to Individualize Instruction and Promote Learning," *Middle School Journal 37*, no. 4 (2006): 44-49.

Lamon, Susan J. *Teaching Fractions and Ratios for Understanding: Essential Content Knowledge and Instructional Strategies for Teachers*. New York: Routledge, 2012.

Larson, Matthew R., Francis Fennell, Thomasenia Lott Adams, Juli K. Dixon, Beth McCord Kobett, and Jonathan A. Wray. *Common Core Mathematics in a PLC at Work, Grades 3–5*. Bloomington, IN: Solution Tree Press, 2012.

Moss, Connie M. and Susan M. Brookhart. *Learning Targets: Helping Students Aim for Understanding in Today's Lesson*. Alexandria, VA: ASCD, 2012.

National Council of Teachers of Mathematics Commission on Standards for School Mathematics. *Curriculum and Evaluation Standards for School Mathematics.* Reston, VA: National Council of Teachers of Mathematics, 1989. http://www.nctm.org/uploadedFiles/Math_Standards/12752_exec_pssm.pdf

National Council of Teachers of Mathematics Commission on Teaching Standards for School Mathematics *Professional Teaching Standards for Mathematics.* Reston, VA: National Council of Teachers of Mathematics, 2000. http://www.nctm.org/uploadedFiles/Math_Standards/12752_exec_pssm.pdf

Paul, Richard and Linda Elder. *The Miniature Guide to Critical Thinking Concepts and Tools.* Tomales, CA: Foundation for Critical Thinking, 2008.

Scriven, Michael and Richard Paul. *Critical Thinking as Defined by the National Council for Excellence in Critical Thinking.* 1987 Conference paper.

Siegler, Robert, Thomas Carpenter, Francis Fennell, David Geary, James Lewis, Yuraki Okamoto, Laurie Thompson, and Jonathan Wray. *Developing Effective Fractions Instruction for Kindergarten through 8th Grade.* Washington, DC: National Center for Educational Evaluation and Regional Assistance, Institute of Educational Sciences, U.S. Department of Education, 2010. http://ies.ed.gov/ncee/wwc/pdf/practice_guides/fractions_pg_093010.pdf

Thompson, Tony and Ron Preston. "Measurement in the Middle Grades: Insights from NAEP and TIMSS," *Mathematics Teaching in the Middle School 9,* no. 9 (2004): 514–519.

Tunstall, Pat. "Teacher Feedback to Young Children in Formative Assessment: A Topology. *British Educational Research Journal 22* (1996): 389–395.

Van de Walle, John A., Karen Karp, and Jennifer Bay-Williams. *Elementary and Middle School Mathematics: Teaching Developmentally.* Boston, MA: Allyn and Bacon, 2012.

Van de Walle, John A., Karen Karp, Lou Ann H. Lovin, and Jennifer Bay-Williams. *Teaching Student-Centered Mathematics: Developmentally Appropriate Instruction for Grades 3-5.* New York: Pearson, 2013.

About Compass Publishing

Compass is the educational books imprint of publisher Brigantine Media. Materials created by real education practitioners are the hallmark of Compass books.

For more information, please contact:

Neil Raphel

Brigantine Media | 211 North Avenue | Saint Johnsbury, Vermont | 05819
Phone: 802-751-8802
E-mail: neil@brigantinemedia.com | Website: www.compasspublishing.org

ORDERING INFORMATION

Quantity Sales

Special discounts for schools are available for quantity purchases of physical books and digital downloads. For information, contact Brigantine Media at the address shown above or visit www.compasspublishing.org.

Individual Sales

Brigantine Media/Compass Publishing publications are available through most bookstores. They can also be ordered directly from Brigantine Media.
Phone: 802-751-8802 | Fax: 802-751-8804
www.compasspublishing.org

www.ingramcontent.com/pod-product-compliance
Lightning Source LLC
Chambersburg PA
CBHW080246170426
43192CB00014BA/2585